Evolutionary Biology – New Perspectives on Its Development

Volume 4

Evolutionary biology has been a remarkably dynamic area since its foundation. Its true complexity, however, has been concealed in the last 50 years under an assumed opposition between the "Extended Evolutionary Synthesis" and an "Alternative to the Evolutionary Synthesis". This multidisciplinary book series aims to move beyond the notion that the development of evolutionary biology is structured around a lasting tension between a Darwinian tradition and a non-Darwinian tradition, once dominated by categories like Darwinian Revolution, Eclipse of Darwinism, Evolutionary Synthesis, and Post-Synthetic Developments.

The monographs and edited volumes of the series propose an alternative to this traditional outlook with the explicit aim of fostering new thinking habits about evolutionary biology, a multifaceted area composed of changing and interacting research entities and explanatory levels. Contributions by biologists and historians/philosophers are welcomed. Topics covered in the series span from (among many other possibilities):

- An Overview of Neutralist Theories in Evolutionary Biology
- Developmental Biology: From Reductionism to Holism and Back
- Selection Theories Beyond Hard and Soft Inheritance
- Divergent, Parallel, and Reticulate Evolution: Competing or Complementary Research Programs?
- The Rise of Molecular Biology: Between Darwinian and Non-Darwinian
- Biologizing Paleontology: A Tradition with Deep Historical Roots
- The Darwinian Revolution and the Eclipse of Darwinism: Blurring the Historiographical Lines
- Darwinism, Lamarckism, Orthogenesis: Can We Really Define Them by Their Hard Explanatory Cores?
- The Evolutionary Synthesis: A Fabricated Concept?
- The Opposition to the Evolutionary Synthesis: Criticizing a Phantom?
- A Reversed Perspective: Approaching Charles Darwin from the Pre-1859 Period
- The Long Development of the Multilevel Paradigm in Evolutionary Biology
- Self-Organization: A Research Tradition from Morphology to Cosmology
- Human Evolution: Sociobiological or Sociocultural?

More information about this series at http://www.springer.com/series/16175

Thomas E. Dickins

The Modern Synthesis

Evolution and the Organization of Information

Springer

Thomas E. Dickins
Department of Psychology
Middlesex University
London, UK

ISSN 2524-7751 ISSN 2524-776X (electronic)
Evolutionary Biology – New Perspectives on Its Development
ISBN 978-3-030-86421-7 ISBN 978-3-030-86422-4 (eBook)
https://doi.org/10.1007/978-3-030-86422-4

Cover photo by GLady is licensed under CC0
https://pixabay.com/photos/mosaic-fish-tile-art-ceramic-200864

This Springer imprint is published by the registered company Springer Nature Switzerland AG.
The registered company address is: Gewerbestrasse 11, 6330 Cham, Switzerland

To David, Anne, and Ben, for their love and inspiration.

To Nicola, Jack, and Isabella, for their unwavering love and support.

Preface

For the last 15 years I have been responding to recent critical analyses of the Modern Synthesis in evolutionary biology. These analyses take several forms, but many of them focus upon the role of development in evolution with some claiming that the Modern Synthesis neglected development during its formation. This neglect has led, some argue, to a focus upon population thinking and to giving the gene causal priority in explanations. This complaint is closely aligned with the latter stages of the synthesis, a period critics associate with adaptationism and a focus upon natural selection as the main cause of evolution. The solution is to incorporate development and to build a mechanistic theory of the emergence of form as part of a renewed evolutionary theory. In this way, the Modern Synthesis is to be extended to incorporate new phenomena but also new explanatory structures. However, while those seeking extension rely upon the most recent of biological findings to illuminate core phenomena, their explanatory structures owe much to historic approaches to evolution grounded in transformational thinking. This observation does not amount to a criticism, because it is always possible to throw the baby out with the bathwater. But it is to note that the proposed Extended Evolutionary Synthesis is to be created by incorporating developmental phenomena as well as theories that have at least a part of their origins in pre-Darwinian thought. The extension reaches into the past and the present from a point somewhere in the middle of the twentieth century.

Springer's book series *Evolutionary Biology: New Perspectives on Its Development*—takes a multidisciplinary view of evolutionary biology and aims to move beyond the dichotomous tension between a modern and an extended evolutionary synthesis, between Darwinian and non-Darwinian modes of thought. The series aims to bring new thinking to well-worn debates. During my own production of papers and reviews in response to specific claims from advocates of an Extended Evolutionary Synthesis I have realized just how complex, subtle, and dynamic evolutionary theory in fact is. While I am committed to a version of the Modern Synthesis, I felt that my particular version of it needed scrutiny and full exposure, as did the core assumptions at the heart of my criticism of extension. To that end I decided to write this book-length treatment.

I describe this book as an Easter Egg. The outer shell, which is the initial presentation the reader meets, is a negative thesis: I do not think any of the proposals for an extended synthesis work in their own terms for analytic reasons that I give. Of course, in doing this I am aware that I am presenting my own axiomatic take on things, but my hope is that in explicitly exposing that it will at least help to clarify the mechanics of the debate. But inside the Easter Egg is a positive thesis, which I believe is a new perspective in keeping with the ambitions of the book series. I think that a principal source of disagreement between those defending the Modern Synthesis and those advocating for extension has been around the concept of information. *Information* has been persistently used to make claims for and about the Modern Synthesis, most especially with reference to genes and developmental programs, and this usage has been loose and colloquial. As a result, a semantic view of information, of information as something contentful and out in the world to be harvested, to deliver instruction, has come to be the default interpretation of information. Information has been objectified, and this is a mistake. Information is instead a functional relationship between an input and a system, between data and context. This view sees information as something realized by systems. In this sense, information is the consequence of the proximate operations of biological systems, and evolution organizes those systems, and hence organizes information. I think this view of information is embedded within the Modern Synthesis, and I demonstrate this by applying it to relevant arguments throughout the book. But it is also a view closely aligned with aspects of developmental systems theory, which has strongly influenced the movement for extension of the Modern Synthesis. My belief is that acceptance of this view of information will lead to a different understanding of the Modern Synthesis, and remove the critical claims based around gene centrism.

In the final chapter, I draw all of this together and make comment on the theoretical aspirations of those pursuing the Modern Synthesis and those seeking extension. In my closing comments, I claim that the Modern Synthesis is a framework theory, founded in an axiomatic structure. As a framework it acts to corral non-axiomatic theories of development, and other phenomena, by setting the parameters for their explanations. This structure is one delivered in the service of an idea, and that idea has been the unification of biology. Presently, I see no reason to abandon it.

London, UK Thomas E. Dickins

Acknowledgments

Many people have discussed the ideas in this book with me. I am indebted to Kevin Laland, John Odling-Smee, and Tobias Uller for discussion and formal debate around the concept of an extended evolutionary synthesis at points over the last 15 years. While I find myself in opposition to their views, I have benefitted from listening to their advocacy, as well as from reading their work. Ben Dickins (my brother), Edgar Porcher, Max Steuer, Paul Taylor, Richard Webb, and Andy Wells have all read and critically discussed draft chapters during our weekly meetings convened on Zoom during the COVID-19 pandemic, but under the auspices of the London School of Economics (LSE). This was a tremendous service, and I am thankful for their patience and guidance. The Centre for the Philosophy of Natural and Social Sciences at the LSE has supported this project by continuing my honorary position as a Research Associate. This role has led to many useful conversations over the years, and Helena Cronin must be thanked for bringing me into this creative, theoretical fold. Oliver Curry and Qazi Rahman are still partaking in a long conversation with me about how to think about biology, and their insights and different ways of thinking have proved important stimuli for my own development. Daniel Nettle has had a huge impact on my thinking, and I am particularly grateful for his endurance and inputs on a walk in 2021, when I talked incessantly about evolutionary developmental biology. Occasional email correspondence with Louise Barrett, David Haig, Douglas Futuyma, Thom Scott-Phillips, David Shuker, Betty Smocovitis, Erik Svensson, and Stu West over the last few years has enhanced my thinking and highlighted errors too. They have this in common with the members of the Behavioral Science Lab at Middlesex University: Claire Gellard, Warren Horrod-Wilson, Amanda Mead, Kirsty Neller, Sabrina Schalz, and Jonathan Sigger. The editor-in-chief of the book series into which this falls, Richard Delisle, has been enormously supportive and encouraging as have the team at Springer.

My father, David Dickins, provided a critical read-through of the book. His zoological and psychological insights have been as useful as ever, and it is his interests that first caught my attention as a boy. Similarly, my mother, Anne Dickins, has continuously dropped adroit, geologically inspired comments throughout my education and career. With parents such as these I was canalized toward evolutionary theory, as was my brother. Finally, I could not have considered writing this book without the support of my wife, Nicola, and the enthusiasms of my children, Jack and Isabella. Nicola has supported an odd working regime, obsessive habits, and my near constant disappearance to "just add something" in my office all the while patiently bringing tea and coffee. My children's interests in natural history have drawn me into nature to watch and learn. And that is really what matters.

Contents

List of Figures

List of Tables

Introduction

1

Abstract

This book is about claims that the Modern Synthesis requires an extension. In this chapter, I outline the approach taken to inspecting this position and give a brief overview of the reasons some have come to argue for an extension. In doing this, I introduce core elements of the argument to come, including the distinction between theories of population and of form and the concept of information. Population-level theories are those of Darwinism and the Modern Synthesis, which distinctly removed transformational arguments of the sort associated with Lamarck. Information has been recruited by many theorists, but its use has been colloquial, and this has led to problematic interpretations. It is a plea for the proper use of information that is at the heart of this book, and in this chapter, I signal its importance. This chapter, and this book, do not exhaustively inspect all aspects of the argument for the extension, but predominantly focus upon claims made from development. This choice has been made to demonstrate the kind of thinking at play on both sides of the debate.

Keywords

Conceptual ecology · Modern Synthesis · Extended Evolutionary Synthesis · Development · Population · Form · General Darwinian Theory · Special Theory of Evolution · Theoretical structure · Information

1.1 Scholasticism and Conceptual Ecology

While writing this book, I had a useful conversation with the psychologist Richard Bentall and my father, David Dickins, about other matters. During that conversation, Bentall mentioned a theological difference between Christianity and Islam. He argued that the Islamic tradition regards the Koran as the given word of God; the Christian tradition sees the Bible as an interpretation of the word of God. By

T. E. Dickins, *The Modern Synthesis*, Evolutionary Biology – New Perspectives on Its Development 4, https://doi.org/10.1007/978-3-030-86422-4_1

inference, this means that Christian theologians and scholars have a theoretical job to do, while Islamic scholars have an educational mission.

This caused me to wonder whether advocates of particular scientific positions, paradigms, and theories operate in one or both of those modes—the interpretive or the pedagogic. This book is about the Modern Synthesis in evolutionary biology and those who have, in recent years, sought to extend it. That use of the term *extend* is broad, with some advocates of an extended thesis looking for minor changes to what is seen as the standard theory of evolution and others looking to replace it completely. Unlike Islam and Christianity, there is no foundational text for the Modern Synthesis, but rather a collection of texts that grew from interpretations of Darwin's *Origin* and the responses from scientists to it. This in turn led to a vast empirical and theoretical literature exploring the hypothesis space that is associated with the Modern Synthesis. Empirical and theoretical work necessarily makes an interpretation, but quite often, that interpretation is a declarative statement along the lines that *the Modern Synthesis causes us to think x, here are some entailments of x empirically or theoretically endowed.*

There is also a meta-literature within the history and philosophy of science, which I see as a discipline positioned on a continuum between pure historians and pure philosophers. In the middle are those who use the historical example to draw out philosophical points, which runs the risk of sampling error and should not be treated inductively, but can be enormously revealing about the array of thoughts possible within a scientific realm. These scholars also act to interpret the words of evolutionary biologists and to make sense of the Modern Synthesis.

All of the preceding scholars, then, are interpretive and also exploratory. While the scientific response is perhaps a more direct interaction with and interpretation of Nature, the meta-literature is informed by science and in turn informs scientists, helping them to consider and adjust their own view. Moreover, there are scholars who operate at the direct and meta-levels. All, then, share the project of interpreting Nature at varying degrees of separation.

This interpretive space is dynamic as new findings and new ideas continue to enter. It would be strange to find the interpretations and journeys of all scholars in perfect alignment with one another. As they diverge, we should expect some to note their parting of ways with particular positions and thinkers and to make clear the distance between them and the directions they are individually taking. A key question in the current debates about the Modern Synthesis is whether or not all of the scholars addressed to its interpretation are roaming the same land. If they are, then perhaps they are mapping new regions and occasionally advocating additions to current global descriptions. This might be equivalent to crossing England from West to East. If your life had been lived only in the county of Devon, your general model of England would currently be one of extensive dairy pasture and the depauperate ecology that sustains. A trip to the county of Essex will expose you to vast deserts of cereal crops and a different but not completely different ecology.[1] One should not be an essentialist about England, but that area which we choose to call England varies in discernible ways, and one can map ecological gradients across it. This journey would impose a certain generality on account of our Devonian ecologist.

Continuing the metaphor of the theoretician as ecologist, mapping a space of possibilities, we might think of another Devonian who had never traveled. Her data on diversity and abundance will be mapped over time, and she will have a phenomenally detailed account of Devon in terms of its flora, fauna, and their interactions. She will be a specialist in that community, and the granularity of her accounts will be finer than those of her traveling colleagues.

If we think of England as the whole of evolutionary biology, then we have a country made up of different counties. Each county would be a particular specialism, akin to population genetics, the study of speciation, paleobiology, and so on. And within each county, traditions would develop to deliver those tasks, and the local management of ideas would take on a particular pattern that delivered local scientific understanding but also helped to make sense of the practices in neighboring counties. But the further east one traveled from Devon, the more stretched your local ecological model would become.

We can stop the metaphor at this point and draw some lessons. First, evolutionary biology, while about evolution as a kind of change, makes contact with the whole of biology as a discipline. There is great diversity in its offer. Conceptually that diversity was ever present, but ecologists had to discover it as they journeyed, much as our travelling Devonian discovered new communities on her way to Essex. Nature had to be interpreted. As she traveled, she would meet resident ecologists in each county, and their specialisms would grade across the English landscape, but those from Hertfordshire would have greater interpretative difficulty when talking to those from Somerset than would our Devonian. Nonetheless, the realization of all these ecologies is a wondrous thing, and our itinerant ecologist will make great effort to bring these ideas together, to find a *lingua franca* in which they could all pose their questions, express their answers and move forward together, sharing their collective access to Nature. She will seek unity through interpretation.

In the beginning was the *Origin*, and the *Origin* asserted evolution as a fact but also gave a theory of it. That theory was immediately interpreted in different ways, leading to diversity of opinion. The Modern Synthesis was a period in which that diversity was directly addressed, and a common framework was developed in the spirit of unification. Those seeking an extension, by way of minor modifications to the Modern Synthesis, are in many ways pursuing the same project. They understand the Modern Synthesis as a continuing journey, a constant effort to understand, and they simply look to extend our shared vocabulary. But those seeking to do away with the Modern Synthesis are perhaps those who have stayed firmly within a particular

[1] For example, the Devonian would be surprised to find some, but not enough, Yellowhammers (*Emberiza citrinella*) in Essex that hang on due to available winter seed and some remnant hedgerow for nesting (Bradbury and Stoate 2000).

sub-discipline, landlocked in their county and developing bespoke accounts of their idiosyncratic niche. This has made communication difficult at the national level, and their solution is to start again and create a new *lingua franca*, based on their language, based on their direct access to Nature. This book focuses upon developmental theorists who, for whatever reason, have felt themselves isolated from the Modern Synthesis and in response have presented what they consider to be a different view of evolution. At times this work appears interpretive, and at others it appears pedagogical.

1.1.1 A Survey of the Conceptual Landscape

I regard the Modern Synthesis as two things. First, it is an extended period of activity within the field of evolutionary biology. This was the time in which evolutionary biology was established as a discipline in its own right. Second, the Modern Synthesis is a framework for setting questions and developing explanations. It is a theory but in the grand sense of that term. It represents a way of thinking about biology and evolution that provides scientific understanding. But as a grand theory, or framework, it is not unspecialized or unfocused.

The Modern Synthesis is so-called, thanks to Julian Huxley (2010), because of the resolution of debates between the Darwinian biometricians and the Mendelian mutationists (Chaps. 2 and 3). That resolution was due to the work of early population geneticists who showed a way for particulate genes to engage in a gradual Darwinian process. But the synthesis did not stop at this point, and much else was developed following these leaps in understanding in the 1900s. Indeed, the synthesis was a continued process of unification into the second half of the twentieth century (Delisle 2017; Mayr 1982; Smocovitis 1992, 1996), as we shall see in Chap. 3.

Aspects of the Modern Synthesis have received criticism. Criticism was integral to the process of the synthesis itself, but in the post-synthesis period scientists did not have to wait long for theorists and biologists to start rethinking what the Modern Synthesis delivered and the wisdom of some of its theoretical directions (e.g., Eldredge 1971; Eldredge and Gould 1972; Gould 1982, 2002; Gould and Eldredge 1977; Gould and Lewontin 1979; Vrba and Gould 1986). Some of these discussions were about fine details of the theory. For example, the discussions about punctuated equilibrium that were initiated in the 1970s by Eldredge and then Eldredge and Gould, were really only discussions about the role of allopatric speciation in evolution that had already been had by Mayr (1989). As Mayr noted, the discussions of the 1970s had a tendency to gloss details, interpreting Darwinian gradualism as even in tempo, when in fact Darwin had taken pains to discuss different rates of evolution. But what is most interesting about Mayr's analysis of punctuated equilibrium is how carefully he unpicks all the claims, relates them to prior argument or findings, and points to problems that remain to be resolved. He is also clear to separate this discussion from saltationism, which was the idea of the spontaneous macromutations upheld by early Mendelians.

What Mayr's fastidiousness tells us is that there are debates to be had within the Modern Synthesis that are considered reasonable, and the framework gives the limits. Those limits are marked by topics that are regarded as either anathema to the Modern Synthetic way of thinking or as having been resolved. Gould (2002) famously discussed these limits as having been established first through a process of restriction in which Lamarckism, saltationism, and orthogenesis were removed from evolutionary thinking. But as I have noted elsewhere with Ben Dickins, this restriction still permitted a good amount of pluralism in thought and is perhaps better associated with the neo-Darwinian period (Dickins and Dickins 2018). Gould's second stage of the synthesis was one of a hardening, of a certain form of dogmatism taking hold that focused upon natural selection as the predominant source of evolutionary change. But as Dickins and Dickins show, at least one leading textbook of evolutionary biology lists the sources of evolutionary change as natural selection, gene flow and genetic drift (see Futuyma 1986; Ridley 2004).

1.1.1.1 Some Comments on Extending the Synthesis

In the introduction to their edited book on the *Extended Evolutionary Synthesis*, Pigliucci and Müller tell us that the Modern Synthesis has defined evolutionary theory since the 1940s even in light of developments in methods and domains such as molecular genetics, developmental biology, and the various -omics (Pigliucci and Müller 2010a). They tell us that they have collected leading experts to write under the title of an *Extended Synthesis* in order to display what they see as a "multifaceted research program" that is a "work-in-progress" (Pigliucci and Müller 2010a: 3). This research program has emerged from these new developments.

> Whenever we talk to colleagues who are inclined toward a conservative position about the status of evolutionary theory, we are confronted with the question "So, what exactly is so new that we may speak of an Extended Synthesis?" This volume is the beginning of a response to that question... The commonest reaction to our explanations is something along the lines of "But that is already understood as part of the Modern Synthesis anyway." We beg to differ. Many of the empirical findings and ideas discussed in this volume are simply too recent and distinct from the framework of the MS to be reasonably attributed to it without falling into blatant anachronism. Concepts such as evolvability for instance, did not exist in the literature before the early 1990s; phenotypic plasticity was known, but consistently rejected as a source of nuisance, not of significant micro- and macro-evolutionary change. Or consider EvoDevo, an entirely new field of evolutionary research that has emerged in full only since the late 1980s, precisely because of the perceived explanatory deficits of the MS in the realm of phenotypic evolution. Yet another common retort to our arguments is that the new ideas are "not inconsistent" with the framework of the Modern Synthesis; this may very well be true—and most of us would gladly agree—but being consistent with the MS is not at all the same thing as being a part of the MS! (Pigliucci and Müller 2010a: 4)

Pigliucci and Müller hypothesize that biologists may have simply upgraded their own understanding of evolutionary theory without reference to the distance traveled from the concepts in play during the formation of the Modern Synthesis; and indeed, many of them may never have had the time to do the historical work in their own field. The implication is that biologists are enculturated by the science they are taught and do, and the label of the Modern Synthesis has been applied without much thought to their own practices. Their own practices may well be entirely consistent and rational, but they are not necessarily those of the synthesis. What this means is that Pigliucci and Müller think that the Modern Synthesis represents a clear and defined set of theoretical assumptions and ideas and that those have been challenged by work since the 1940s.

After making these points, Piggliucci and Müller move on to a sketch of the Modern Synthesis that covers ground I cover in more detail in Chaps. 2 and 3. Indeed, I have written those chapters in order to present the kind of history of evolutionary thought that is most typically referred to by those advocating for an Extended Evolutionary Synthesis, drawing on the same literatures that they do. This helps me to place the arguments I deal with, but it does not mean that *this* history is a correct rendition of this long period in scientific development. It simply shows a set of possible interpretations and ideas. More importantly, it shows what the scholars involved in the ongoing arguments both for and against the Modern Synthesis are arguing about. In this case, Piggliucci and Müller tell us that key areas of biology were left out of the Modern Synthesis, including developmental biology and ecology: this is the claim for isolation, for being left within a county studying its local natural history. What was included in the synthesis they highlight by quoting Futuyma's textbook in which he lists the key principles of the synthesis (Futuyma 1986), given below as a bulleted list:

- Populations contain genetic variation.
- Genetic variation arises at random via mutation and recombination and is not directed toward adaptation.
- Populations evolve by changes in gene frequency themselves caused by genetic drift, gene flow, and especially natural selection.
- Adapted genetic variants have slight phenotypic effects making phenotypic change gradual (although there are some alleles with discrete effects that may be useful).
- Diversification is caused by speciation, normally caused by the gradual emergence of reproductive isolation among populations.
- These micro-evolutionary processes above can, over a long time, lead to macro-evolutionary change.

This list sets the agenda for possible challenges, and indeed all these principles are expressed with a clarity that permits challenge. But Pigliucci and Müller clarify the kind of challenge their book contains, and they do this by giving what they consider to be a parallel example, that of the Central Dogma of Molecular Biology.

> The dogma … stated that the flow of information in biological systems is always one way, from DNA to RNA to proteins. Later on, however, it was discovered that the DNA>RNA flow can be reversed by the appropriately named process of reverse transcription, which takes place in a variety of organisms, including some viruses and eukaryotes (through retrotransposons). Moreover, we now know that some viruses replicate their RNA directly by means of RNA-dependent RNA polymerases, enzymes also found in eukaryotes, where

> they mediate RNA silencing. Prions have shown us how some proteins can catalyze conformational changes in similar proteins, a phenomenon that is not a case of replication, but certainly qualifies as information transfer. Finally, we also have examples of direct DNA translation to protein in cell-free experimental systems in the presence of ribosomes but not of mRNA. All of these molecular processes clearly demolish the alleged central dogma, and yet do not call for the rejection of any of the empirical discoveries or conceptual advances made in molecular biology since the 1950s. Similarly, we argue individual tenets of the Modern Synthesis can be modified, or even rejected, without generating a fundamental crisis in the structure of evolutionary theory—just as the Modern Synthesis itself improved upon but did not cause the rejection of either Darwinism or neo-Darwinism. (Pigliucci and Müller 2010a: 9–10)

This is an odd rendition of the Central Dogma that specifically neglects a major 1970 paper by Crick in which he discusses his original 1958 version of the idea and then modifies it (Crick 1970). In its original formulation and its modified version, viral RNA replication, DNA→protein, RNA→DNA, and the role of prions were all discussed and incorporated. What were ruled as unlikely were data flow from protein to RNA or DNA. In Chap. 7 I discuss the Central Dogma and the detail of this 1970 paper in the context of epigenetic processes, an area that a number of advocates of the Extended Synthesis have explored.

In spite of this quibble, I agree with the broader point that we can see the transitions from Darwin to the Modern Synthesis as modifications and that those modifications left intact much of the original work, but not all. And it is certainly the case that these transitions did not lead to crisis, but they did lead to robust disagreement and the formation of distinct interest groups, for example, the biometricians and Mendelians of the early twentieth century, whose debate proved useful (Chap. 3). We might package this as *dissent with modification* and note that this is normal business for any science. And put this way, we get a sense of what Pigliucci and Müller meant about consistency—for them, modification is a process that gradually shifts you from the original form toward something new, an evolutionary process in its own right. At some point, the emerging novelty is consistent with early ideas, while different enough to warrant a title: *what is it we should call this conceptual space?* And as you might expect, Pigliucci and Müller provide a diagram to denote this spatial relationship (Love 2017) and have the Extended Synthesis encapsulating the Modern Synthesis, which in turn encapsulates Darwinian evolution (Pigliucci and Müller 2010a: 11) (Fig. 1.1).

If I am right, then the claim of Pigliucci and Müller is a claim to novelty and they characterize novelty predominantly in terms of phenomena previously unaccounted for by the Modern Synthesis but also in terms of a move away from traditional views of the role of natural selection and the nature of adaptation, inheritance and replication, developmental biology and the informational role of the gene. These shifts, they argue, are a result of novel discovery and inclusion.

Pigliucci and Müller are engaging in a taxonomic exercise. They are listing features and processes that differ between the Modern Synthesis and the conceptual space that has since emerged. They feel ready to declare a new species of theory, the Extended Synthesis, which is historically contingent and thus conceptually related to

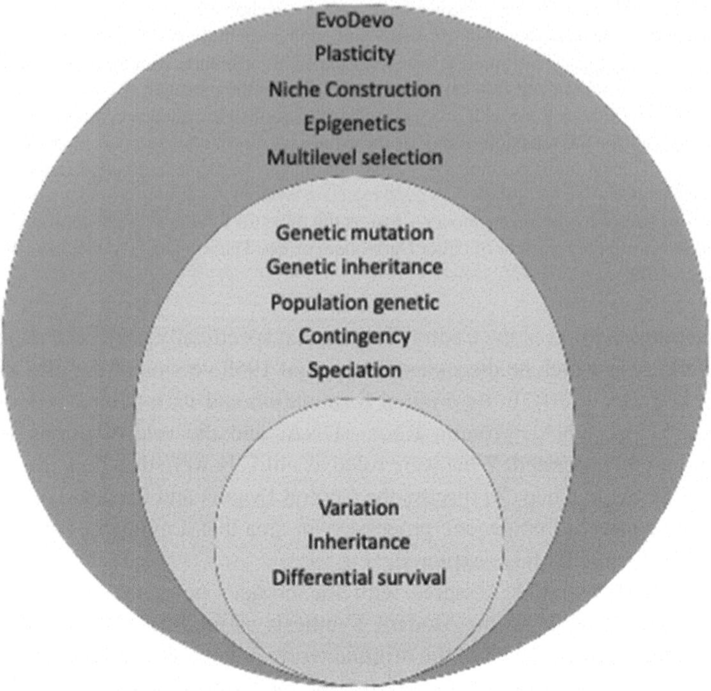

Fig. 1.1 A spatial representation of the Extended Synthesis. A Venn diagram depicting the relationship between components of Darwinian theory (inner circle), the Modern Synthesis (middle circle) and the proposed Extended Synthesis (outer circle), adapted from Pigliucci and Müller (2010a, p. 11). EvoDevo is Evolutionary Developmental Biology

or consistent with the Modern Synthesis, but the two are moving toward reproductive isolation. And yet, using the same data about new ideas in evolutionary biology, emerging from new phenomena, it would be possible for our conceptual ecologist to track a gradient between, for example, those adopting adaptationist programs of research in behavioral ecology which emerged in the final stages of the Modern Synthesis (Chap. 3) and those interested in phenotypic plasticity expressed under various conditions.

In this book, I discuss multi-level selection (Chap. 3), phenotypic plasticity (Chap. 5), evolutionary developmental biology (Chap. 6), epigenetics (Chap. 7) and niche construction theory (Chap. 8) as these areas have provided core phenomenological claims for advocates of the Extended Synthesis, and feature heavily in the major book length treatments of this advocacy (Huneman and Walsh 2017; Pigliucci and Müller 2010b; Uller and Laland 2019a). What is notable about these areas is how they can be lined up on a gradient organized from a focus upon *population* thinking to think about *form*.

Darwin is attributed with the introduction of population thinking, making evolution a kind of change that happened to populations and not to individuals. In doing this, he stepped away from a transformationist notion of development and evolution as closely related, if not identical processes. Lamarck's commitment to the

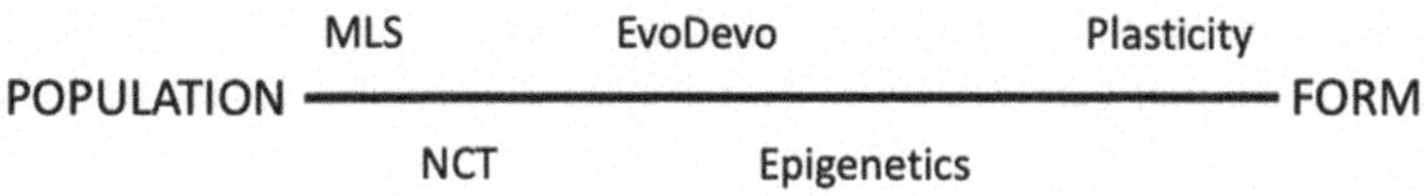

Fig. 1.2 The population-form continuum. One hypothesis about how the core areas recruited by those advocating an Extended Synthesis can be distributed on a continuum of concerns from population level accounts of evolution, to those focused upon the emergence of form. MLS is Multilevel Selection Theory, NCT is Niche Construction Theory, and EvoDevo is Evolutionary Developmental Biology

inheritance of acquired characters was to a developmental process at the individual level, mediated by behavior to a large extent, that fitted the form of an organism to the environment to solve crucial problems. This was his notion of adaptation, and he was metaphysically committed to such changes being heritable. Evolution was the change in form within species over time as a result of such processes. For Darwin, the population changed due to the differential success of variants in the struggle for life. Variants were of form, and his struggle was to understand the sources of variation, as well as the mechanisms of inheritance, but this clearly posed a different problem to that considered by Lamarck. Darwinian evolution was not about changes in form but rather how changes in form could lead to changes in populations. Ultimately, those populations could be characterized by the predominant forms present and those forms, duly selected, were adaptations.

The arrival of neo-Darwinism and then the Modern Synthesis gave a mechanism of inheritance that enhanced Darwin's view. And in so doing it allowed a new direction of gaze, from the gene up. The gene was clearly a crucial element for development *and* for the evolutionary process and thus had a role to play in the emergence of form ontogenetically and its ordering phylogenetically. This direction would feel familiar to Lamarckian thinkers, but the difference was that Darwinians had their eye always upon population change.

In Fig. 1.2, I have depicted one possible hypothesis about how the central themes used by advocates of an Extended Synthesis might relate to one another along a continuum from *population*-level accounts to accounts focused upon the emergence of *form*.[2] I take the population end of the continuum to be traditionally Darwinian and also the focus of the Modern Synthesis. As a result, I have positioned Multi-level Selection Theory close to that pole. Niche Construction Theory, with its focus on the effect of organism activity upon population-level selection dynamics, is also toward that end of the continuum. But these theorists also discuss the role of agents in the creation of their own form. Evolutionary Developmental Biology sits in the middle with a clear interest in the role of the development of form upon selection dynamics at the population level. Epigenetics is closely related to this discipline, but its principal role is in gene regulation and thus the expression of form. Nonetheless, epigenetic transmission has been recruited to selection-based arguments by some. Finally, plasticity is all about the reactive nature of form with relation to the environment, and while a part of the developmental response, it is often recruited

[2] A related idea about a transition from theories of form to theories of genes has been mooted in the literature (Pigliucci 2007), and I shall return to this idea in Sect. 1.1.1.3.

across arguments for the Extended Synthesis, in part as a counter to gene centrism and notions of formal determination. This is why it occupies a place near the *form* pole.

What Fig. 1.2 does not yield is an account of the direction of travel for those engaged within each of the core areas. It is presented as a possible gradient based on the observations of this conceptual ecologist. Using this continuum, it is possible to come to a phenomenon such as a plasticity and ask whether this capacity has been selected. Answering this question would not preclude further exploration of the impact of plastic responses upon selection dynamics. Indeed, that would inevitably become a part of that account. Similar questions could be addressed within epigenetics. Some may be surprised to see me separate Evolutionary Developmental Biology from plasticity and epigenetics, after all, these are mechanisms involved in development. This is true, but there is more to development than this, and a key argument from Evolutionary Developmental Biology has been around the conserved, compartmentalized nature of development and what this might mean for evolution. This is why I have positioned this field slightly left of center in the figure. Nonetheless, it is still possible to arrive at the facts of compartmentalized development and ask whether this trait has been selected.

From left to right is the direction of travel for those applying the Modern Synthesis (Fig. 1.2). My suspicion is that many seeking an extension of the synthesis start their journey somewhere toward the *form* end of the continuum and head toward the other pole, in much the way Darwin had to when confronted by the diversity of life. Where Darwin invented this for himself, modern scholars are armed with Darwinian theory and know to look for mechanisms of inheritance and the production of variation. As they travel leftwards, they reveal various localized levels of mechanism that produce form and sit well above the level of the gene. Naturally, they wonder whether or not these levels are both necessary and sufficient for the production of *form*, and because there is variation at these levels, they wonder if some kind of general Darwinian process might unfold (Sect. 1.1.1.2, Webb 2011). For this, they would need inheritance, and within the epigenetic and niche construction literatures there is much discussion of non-genetic inheritance and how that might enable evolution.

This use of Fig. 1.2 is highly rhetorical. I cannot hope to truly fathom the real starting points and motivations of each and every one of those advancing arguments for an extension of the Modern Synthesis. But my hope is that it makes for a useful tool to at least pry open the discussion and delve into the mechanics of the arguments. For example, thinking about travel from *form* toward *population* accounts, I can see how readily a scholar might stop and read gene-level theorists and wonder how on earth they truly believe that the data contained within a gene can account for the diversity and complexity they see from their perspective. Gene-level theorists might easily appear to be utterly detached and esoteric in their accounts. But just as they accuse others of privileging the gene, and gene centrism, they effectively privilege developmental processes, sidelining a core element of continuity. This is something often excused by claiming that the Modern Synthesis deliberately left development out of the unification of biology and arguing that it is now time to prioritize these processes and to see if they do more than explain ontogeny (Chaps. 4 and 6).

Figure 1.2 is no more rhetorical than Fig. 1.1. Pigliucci and Müller's (2010a) placing of multi-level selection theory and niche construction within the extended synthetic set conveniently ignores the fact that group selection ideas occurred within the period of the Modern Synthesis (Wynne-Edwards 1962) and that Darwin discussed the core concepts of niche construction in his work. Moreover, plasticity has received considerable theoretical attention from scholars within the Modern Synthesis. I suspect some of these oversights are also a consequence of traveling from the *form* end of the continuum and arriving at these ideas before fully immersing oneself in the role of the gene and *population* thinking. But the inclusion of multi-level selection is puzzling as this relies upon Darwinian selection, fitness and population-level change. This is clearly a Darwinian view at the base, and proponents such as Wilson recruit single gene models in keeping with the phenotypic gambit from behavioral ecology, a product of the Modern Synthesis (Grafen 1984; Wilson 1975). All told, this does not appear to be a radical departure from standard theoretical thinking.

One reason that some may choose to start with *form* is that they wish to counter claims that natural selection is the preeminent creative force in evolution. Developmental processes can be invoked by those proposing an extension in order to discuss the potential limits on available variation in a population (Lewens 2019). Lewens sees this as a mild claim that does not particularly challenge the Modern Synthetic view that natural selection underlies the emergence of adaptation. Indeed, a clear relationship between an underlying selection dynamic and available variation constrained by developmental factors would make for what Lewens' terms an *enriched* account of a given phenomenon. But Lewens also points to a stronger reading of the view that development is involved in the creation of adapted form. This rests on processes like learning that enable an adaptive fit during ontogeny. Lewens is quick to point out that while the precise outcome of a learning event might not be something selected in ancestral conditions, the capacity to learn may well be so, and in so doing, he is in keeping with standard theoretic views of plasticity (Chap. 5).

In making this last point, Lewens gives an imaginary example of an animal learning to avoid a novel predator. This learning is labelled adaptive, providing an adaptive fit, and my reading of his paper is that this is supposed to capture the idea of adaptation; thus *adaptive* is used as a synonym for *adapted*. I think this use of adaptive, which is common on both sides of the debate for simple reasons of common usage, is potentially perilous. Adaptive means given to or prone to adaptation, which in turn is about adapting, or modifying to a new purpose or adjusting to new conditions (Stevenson 2010). This notion of modification and adjusting is another view of plasticity. Within evolutionary biology, we can talk of populations adapting, and more generally, of populations being prone to adaptation, where adaptation is a technical term. Populations are then adaptive or plastic, and under Darwin and the Modern Synthesis, they are the proper focus of attention. But, while it is true that, for example, behavior enables organisms to adjust and modify in light of new conditions, this does not carry with it any technical evolutionary connotations—or rather, it should not until extra conceptual work has been done. In evolutionary terms, where we regard a species trait as an adaptation, we regard it as a trait with a positive fitness function. Not all behavioral responses will contribute to average lifetime inclusive fitness, and Williams advised us, in his contribution to the Modern Synthesis, to take the work of identifying adaptations seriously and cautiously (Williams 1996).

To return to Lewens' example, the animal learns to avoid a novel predator, most likely because core indices of predation threat are present in that animal. Being able to use those indices as data in an appropriate learning response is a credible adaptation that allows adjustments in the present by the potential prey item. In strict evolutionary terms, in keeping with the Modern Synthesis, the adaptation is the learning mechanism, and the particular *token* behavior that releases the animal from threat is not. However, we might be able to say that that *kind* or *type* of behavior is an adaptation, but in so doing, we are in fact describing some of the parameters of the mechanism that delivers it. Most typically, in this latter case, we would describe it as predator evasion behavior. For learning to work, as what Lewens terms an exploratory process, the error will also be produced by these mechanisms, and we have no reason to label errors as evolved adaptations, especially as they may be costly.

I think being clear about these distinctions is important because noting adjustment and modification is not the entirety of noting adaptation in the strict evolutionary sense. Ironically, the looser uses of adaptive enforce a kind of pan-adaptationism upon explanations in the absence of evidence, something which the forebears of the current arguments for extension are worried about (Gould and Lewontin 1979; Gould and Vbra 1982). I strongly suspect that where some see adaptive fit between an organism and the world, they by default assume adaptation and thus evolution. Moves by learning theorists to model learning as a form of selection by consequences may have enhanced this view (Smith 2018). This perspective is again a consequence of traveling from *form* toward *population* on the continuum in Fig. 1.2, while having in mind knowledge of what Darwinian evolution is.

1.1.1.2 Further Abstractions

Webb (2011) has proposed a distinction between a General Theory of Evolution, that applies to "order-generating evolutionary mechanisms you might find anywhere" (p.94) and the Special Theory of Evolution, which is a "particular worked example of the General Theory" (p.94) that he aligns with the Modern Synthesis.

Webb's general theory covers systems that meet six criteria.

1. *Information*. An information base, where information content is associated with differential instantiation.
2. *Variation*. Variance within that information, not just as a starting condition but recurring, by whatever mechanism (such as copying errors) or by introduction from elsewhere.
3. *Recombination*. Some de facto recombination of the information taking place, in which the "shuffling" of informational components is associated in some systematic way with the differential instantiation.
4. *Selection*. Selection against some vaguely persistent criteria, that leads to differential distribution of the information according to its differential instantiation.

5. *Replication*. Some form of copying, or relative increase, or privileging, whereby the information redistributed by selection differentially and systematically changes its frequency in the system.
6. *Iteration*. Multiple repeats of this sequence. As a consequence of this the altered frequencies resulting from replication and selection bias the pool in which variance and recombination are occurring (Webb 2011: 96).

He notes that biological systems enable these criteria, but so can non-biological systems. He gives as an example the high throughput method used in the pharmaceutical industry where chemicals are thrown at bacterial cultures to see if there is a desired effect. The only thing known prior to this process is the formulae of all the chemicals that will be tested. Once an effect is discovered, the scientists look up the formula of the responsible chemical, synthesize lots of variants on this theme, do lots of "mixing and matching" (p.99), transmitting what works to the next generation of tests, and gradually winnow options toward a candidate drug for their target problem.

It is possible that this distinction might make immediate sense of Fig. 1.1, and Pigliucci and Müller's (2010a) notion of consistency between Darwinian theory, the Modern Synthesis and their proposed Extended Synthesis (Dickins and Dickins 2018). While Darwin was focused upon naturally occurring biological systems, he drew heavily upon lessons from animal breeders and also Malthusian ideas about population dynamics. The use of these examples might be seen as reasoning by analogy (Evans 1984; Gildenhuys 2004; Theunissen 2012), and the analogy worked because there is something in this general process that can bring order to systems. It was perhaps less analogy, and more homology. The development of the Modern Synthesis, or the Special Theory of Evolution, was then a detailed effort to understand precisely how naturally occurring biological systems, known as populations, were organized according to these generic principles. Thus, the Modern Synthesis is a consistent extension of Darwinian theory within the biological realm. But what of the proposed extension to that synthesis?

Moving from left to right along the continuum depicted in Fig. 1.2, from *population* concerns to those of *form*, would enable a ready accommodation of all of the phenomena and areas listed. There are available accounts of all of these things that are entirely consistent with the Modern Synthesis, and I attempt to demonstrate this throughout the relevant sections of this book. But for those traveling in the other direction, as noted above, there is a temptation to look to non-genetic sources of variation and inheritance and to argue for evolutionary dynamics as a result. These evolutionary dynamics might be claimed to continue independently of gene-level selection, such as in models of learning that rely on selection by consequences or cultural evolution models that rely upon social learning mechanisms. Indeed, Dawkins, the arch villain of gene centrism, famously made this point about cultural evolution when he mooted the idea of memes as a form of conceptual replicator (Dawkins 1989). Here we might lay claim to the General Theory of Evolution as a possible account of the emergence of order in the relevant cultural systems, but Webb would expect these accounts to conform to his criteria.

We can follow the preceding suggestion a little further. Either scholars declare non-genetic evolutionary processes to be general evolutionary processes that make no contact with the underlying processes that are the business of the Special Theory of Evolution, or they must claim some kind of relationship between General and Special processes. If the claim is for a relationship, that can take the following forms: (i) feedback into selection dynamics or (ii) the narrowing of exploration space.

General processes could establish an order that in turn establishes selection dynamics, which amounts to feedback at the population level. An example of such a claim might rely upon social learning mechanisms delivering particular behavioral outputs in a stable environmental scenario and those outputs constraining aspects of the environment. The lactose tolerance story, so often used by niche construction theorists, is a case in point (Scott-Phillips et al. 2013). Accordingly, humans learned to use milk products as energy and nutrient-rich food sources, and in so doing, increased selection pressures for alleles that did not switch off lactose tolerance in infancy. As a result, adult lactose tolerance increased within certain subpopulations. This form of niche construction has been labelled selective (Uller and Helanterä 2019).

General processes can produce stable phenotypes simply as a consequence of prior *special* processes leading to the selection of adaptations. Those adaptations narrow the phenotypic range at physiological levels (including behavior). Lewens' point about the selection of learning mechanisms to allow learning can fit this category (Lewens 2019). If, for example, learning is a process of selection by consequences, then that learning is a result of those evolved mechanisms, but the selection dynamics that unfold at the learning level are independent of prior selection for the mechanism. This is, in fact only partially true as we should expect constraints on the type of learning and thus the type of thing that can be learned. Nonetheless, behavioral variation can be instantiated. While learning in this context is an individual-level process, we should note that the General Theory approach does not stipulate that systems must be populations.

I am not claiming these two kinds of relationships to be exhaustive. Rather I am pointing out that there is a separation and also possible interaction between general evolutionary and special evolutionary systems. But interaction depends upon the systems working as they do. There is no sense in which the operations and outcomes of cultural learning change how selection can favor particular alleles.

1.1.1.3 Explanation

If advocacy for extension were merely advocacy for recognition of specific interactions that have been noted during investigations into the ontogeny of form, then it would be no more than a claim for additional and interesting topics of study. This is not normally something that requires the endorsement of theoreticians in an enormous and critical literature. Nonetheless, I suspect this is precisely how many biologists read it, and it explains the common "so what?" response that Pigliucci and Müller reference (Sect. 1.1.1.1, Pigliucci and Müller 2010a). As we have seen, these particular authors reduce this to an issue of consistency and argue that consistency does not mean identity, which is of course true. In Sect. 1.1.1.1 I gave some possible interpretations of what is meant by consistency. They were the addition of missing or deliberately evaded, fields of study and the focus upon the emergence or creation of *form* as an analytic starting point. These interpretations, either independently or together, give the impression that advocates of extension see the Modern Synthesis as a categorical statement of interests, and anything additional will have to be an extension. It then becomes a moot point of epistemic arithmetic as to when continued addition leads to a qualitative shift in the scientific enterprise.

Some advocates directly address the idea of qualitative shifts, in the context of paradigm shifts, following Kuhn, and modifications to the outer belt of auxiliary hypotheses, following Lakatos and in so doing raise another aspect of calls for extension: does a different phenomenal focus force change upon the structure of evolutionary explanations (Baedke 2021; Dickins 2020; Otsuka 2019)?

The most recent book length treatment from those seeking extension focused upon the issue of causation in evolutionary biology (Uller and Laland 2019a). This book covered a diverse array of topics, including issues to do with development, plasticity, and niche construction. In my review of the book, I made the following claim, in response to a point from Uller and Laland about the nature of models as abstractions that result from choice of focus (Uller and Laland 2019b):

> The MS is indeed a framework to deliver models. Its core conceit is the definition of evolution given by Darwin, and it has relied upon natural selection to generate hypotheses. But the MS has not been unaware of other ways to produce patterned population change, and drift is always to be considered for example. What natural selection does is enable the construction of falsifiable hypotheses about particular biological systems. As such, the MS might be seen as a viable research program, following Lakatos. As a viable research program, new discoveries about the origin of variation, or competition or inheritance are to be dealt with—do they challenge the hard-core axioms of the program, or can they be accommodated, even with tweaks to the protective belt of day-to-day empirical work? To challenge new predictions must be made. (Dickins 2020: 513)

In this case, Uller and Laland were hoping to emphasize the point that other creative forces are involved in the production of form. I went on to suggest that perhaps the claims for extension could be seen as viable modifications to the protective belt but not as challenges to the core axioms of the Modern Synthesis. In this way, I made sense of the claims for additional phenomena but drew upon Lakatos to make the point that a qualitative shift had not been achieved.

Baedke (2021) has pulled me up on these comments. In his paper he wonders "why, at all, should a critique of evolutionary biology only be considered a valid contender or 'real challenge' if it criticizes (or tries to falsify) the theory of natural selection?" (p.85). Baedke claims that an axiomatic approach to understanding theory change sets the bar too high for critics. I find this response perplexing as I am not entirely sure why anyone should make life easy for critics. Moreover, I was simply responding to arguments from within the volume that directly referenced Lakatos and other discussions of research programs. Thus, Otsuka claimed that the Modern Synthesis is showing signs of degeneration, in keeping with a Lakatosian perspective, and has issued "ad-hoc patches just to save its core doctrines without leading to novel research questions or predictions" (Otsuka 2019: 262). Otsuka's analysis is that the Modern Synthesis is being outcompeted by the rival Extended Evolutionary Synthesis research program. My view of the scene suggested a different use of Lakatos, not least because, as Baedke points out, I see developmental explanations as separate from evolutionary ones such that evolutionary axioms are in some senses impervious to developmental accounts (Chaps. 4 and 6).

Toward the end of the paper, Baedke advocates for pluralism, for the incorporation of developmental accounts into evolutionary theory, and a focus upon desirable features of explanatory power. In doing this, I suspect his journey is from *form* toward *population* (Fig. 1.2) because he does not seem to consider the alternative that development is something to be explained by evolutionary theory rather than made part of it.

In another paper, Baedke and colleagues analyze the nature of explanation at stake in this debate (Baedke et al. 2020). Here they directly discuss Pigliucci's comments about theories of form and genes.

> I will provide a brief conceptual recapitulation of the history of evolutionary theory, with the aim of tracing the broad questions posed by evolutionary biologists at different points in time during the past 150 years, as well as the sort of approaches that have been used to answer such questions. I will then argue that evolutionary theory has shifted from a theory of form to a theory of genes, and that it is now in need again of a comprehensive and updated theory of form. (Pigliucci 2007: 2743)

As Baedke et al. note, this is something that has also been described as a shift from statistical correlation to mechanistic causation (Pigliucci and Müller 2010a) and as something that opposes population thinking. This suggests that Fig. 1.2 at least describes some aspect of the debate, and I am probably right about Baedke's direction of travel. Their view is that differing ideas of what constitutes explanatory power are in fact in play such that advocates of extension, like Pigliucci, are arguing for both complementarity between the Modern Synthesis and an Extended Synthesis, but also for increased explanatory power with the addition of this complement.

Baedke et al. claim that explanatory power "can be assessed by comparing the range of inferences to potentially new counterfactual situations and, accordingly, of answers to w-questions that alternative explanations make possible" (p.7). W-questions are *what if things had been different* questions. As an example, they suggest that an account of a population dynamic might have been made more counterfactually comprehensive with the introduction of developmental factors. They dub such expanded explanations more informative. This relates to the idea of *precision*, which is an outcome of the amount of detail, perhaps from across levels of biological organization, that is incorporated. Baedke et al. note that this is decreased by *idealization*, such as that found in modelling practices (Potochnik 2020), and they further claim this can lead to false answers to counterfactual questions.

"Another criterion for saying that explanations are better argues that they should cite causes *proportional* to their effects, meaning that they have the same degree of precision and thus level of analysis" (Baedke et al. 2020: 9, italics added). An example here might be the developmental claim that relying on genes alone to explain form is not proportional to the effects seen in the emergence of form. Again, this is an aspect of detail. The addition of detail not only facilitates *precision* and *proportionality*, but also *sensitivity* such that more detailed explanations are more sensitive to changes in background conditions and context, and thus less readily generalized.

Baedke et al. recruit these principles to look at explanations from the Modern Synthesis and alternative accounts proposed as a part of an extension. What they immediately note is that both differ with respect to *precision* about what it is to be explained. This is because standard theory "focuses on genes and populations and abstracts away from details on the organismic level" (p.12). By this, they mean that an extended synthesis can "provide information about cultural niche construction, social learning, sources of variability, directionality, causality, modes of inheritance, targets of selection, and tempo of evolution" (p.13). When discussing *idealization,* they use optimal foraging models as highly idealized standard theory models and contrast this with an extended approach that does not assume individuals are rational maximizers. They draw a moot point about distinctions in proportionality between standard and extended approaches.

The entire analysis from Baedke and colleagues is designed to show which explanation is better, a standard, Modern Synthetic one or one from an Extended Synthesis. But they make clear that both camps are in fact trying to explain something different, which makes comparison with regard to efficacy redundant. A question about whether or not a trait is an adaptation is not the same as a question about how that trait develops ontogenetically, nor about the phylogenetic history of its emergence.

Ontogeny and phylogeny can provide useful clues to determining the adapted value of the trait. The use of those clues, and also the error in behavioral signal that can be measured by using optimality models, is all part of the practice of biologists that emerged from the Modern Synthesis. Thus, optimality models are used to idealize the functionality of hypothesized foraging behaviors and to define parameters in terms of inclusive fitness theory. Those models are applied to real animal behavior. The expectation is that actual behavior will be suboptimal, and the next step, for behavioral ecologists, is to develop hypotheses about the causes of that suboptimality. But Baedke et al. complain that this practice uses plasticity incorrectly by seeing it as a further mechanism for fitness maximization. It is true that fitness maximization is a core axiom of inclusive fitness theory, which emerged from the Modern Synthesis (Chap. 3). The addition of plasticity to an optimality model, under this axiom, is the addition of detail to the model, which allows for some sophisticated, and *precise* claims about the functional parameters of the mechanism (Nettle and Bateson 2015). However, it is also the case that behavioral ecologists expect some errors to be chaotic and possibly unavoidable under certain circumstances. Here Baedke et al. would further complain that this approach is externalist and allows no interaction between what is within and what is without in an account of form (Godfrey-Smith 1996). The environment drives it all.

Baedke et al. are probably right that the addition of detail to an explanation can make its prediction more precise, and as a result of that, the probability of very precise outcomes are reduced, making those outcomes more surprising and therefore more informative when they do happen (Chap. 5). But those details have to be the right details for the task at hand. Simply noting that an Extended Synthesis would add more details is not sufficient to endorse this as an explanatory strategy. Evolutionary theory is not about the emergence of form but the selection of form. How form emerges is of no relevance, but that it does is. But, of course, how form interacts with the world impacts upon selection. While it is the case that phenotypic variation is caused by both genetic and other causes during ontogeny, it is the likelihood of those ontogenetic effects being wrought consistently across time that allows evolution. The Modern Synthesis taught us that genes have this property of consistency, whereas ontogenetic exposures do not for reasons to do with the vicissitudes of life but also to do with the nature of individuality. This suggests that genes must be able to cope with a deal of noise in their systemic interactions with the world, but as natural selection shows, this has its limits (Dickins and Dickins 2008). A clear counterfactual that Baedke and colleagues more generally ought to deal with is whether or not evolution could occur without a necessary antecedent cause like the gene? Also, could evolution occur without niche construction, plasticity, and so forth? If the answer to this first question is "no," and to the last "yes," then it becomes clear that niche construction, plasticity, and so forth are not a necessary or sufficient part of evolutionary theory.

1.1.1.4 Information

This brings us to the central theme of this book, and that is information. In Chap. 5, I introduce information and draw a distinction between data and information such that *information* = *data* + *context* (cf. Floridi 2010). This allows me to draw a further distinction between notions of *information as something in the world to be harvested* and *information as a functional relationship between data and context.* It is this last view that I endorse and that I recruit to make sense of the Modern Synthesis and particular interpretations of it by those seeking an extension. Very simply put, organisms do not seek information, but they produce information by processing data that they are designed to respond to. But this is a view that does not just account for organisms but also for their component parts and can be applied within evolutionary and developmental contexts. A part of my claim is that life is information, or rather creates it and that the patterned use of inputs from the world external to a given system demands an explanation in these informational terms. It was this that the Modern Synthesis provided, and it is this question that developmentally oriented advocates of an extension cannot avoid but have in fact tended to. But, as I explain in the book, a large part of this ignorance is due to the colloquial use of *information* as out there to be harvested, and both sides of the debate adopt this usage.

To treat the gene as conveying developmental information is to use information in error. Genes contain data, and that data only has an informational role within the context of a developmental system (Oyama 2000). This idea is present within core writings of the Modern Synthesis, but the distinction between data and information is not drawn and, I claim, has led many to misunderstand the role of the gene in standard theory. The gene centrism of that theory is not a commitment to total developmental determination, to some kind of preformation, but rather to a necessary data input at the start of a complex, multi-systemic response that does deliver form. The concept of the gene that is really at work in the Modern Synthesis is one that enables this necessary input and delivers the constancy across time that it is essential for evolution as a population-level effect. So, the optimality models of foraging theory rely upon this notion of constancy and data in the abstract to enable their hypotheses about behavior, but there is no commitment to the totality of the behavioral response being encoded within a gene, but rather to a complex set of genes catalytically contributing to a long developmental process. Genetic data is at the bottom of a hierarchy of multiple constraints upon the kinds of behavior an organism can produce, not the precise tokens.

The astute reader will note that this use of information is framed as travel in the direction of *form*, from left to right in Fig. 1.2. As such, it is a commitment to statistical correlation, but it is also a commitment to a particular functional account of the mechanistic causes of diversity and form. It provides the explanatory parameters for developmental accounts, and in this way, the Modern Synthesis has always incorporated the phenomena that those advocating extension claim are forgotten. Some describe this as a process of black boxing the true mechanics of development, etc. but as I argue in this book that black boxing or bracketing proved prescient, and the phenomena of development that are presented for the extension are entirely in keeping with the fundamental axioms of the Modern Synthesis; the black boxes were the right size and shape.

1.1.2 Summary

This book represents my interpretation of evolutionary theory and modern debates about its adequacy. I have used multiple sources and doubtless drawn conclusions and inferences that many, including those who consider themselves aligned with the Modern Synthesis, would find puzzling and in error. Those errors are honestly achieved. For the most part, variance in my error will be intrinsically caused, as for everyone. Those intrinsic causes will include intellectual shortcomings but also my learning history over a long exposure.

My professional discipline is behavioral science, and I am especially interested in what behavior is and is for. This has naturally drawn me to evolutionary accounts, and most especially those from behavioral ecology. But I also have a background in history and philosophy of science, and I have long held an interest in just how theories work and are applied. This is another reason for my attraction to evolutionary theory. Finally, I also have some training in ecological methods that I mostly put to use in bird surveys. It was reflecting upon this list of interests and trying to find a common thread that led me to realize I was a *conceptual ecologist* when addressing theory. While trained in the history and philosophy of science, I would argue that my approach to the project of this book, and the prior papers that led to it, was one of surveying, classifying, and organizing. My interest in information theory has many routes, but one is its use in diversity indices in ecological surveys. Here I try to weave information theory into my arguments and adopt it as an organizing principle, although I think it has theoretical potency also.

The book begins by retelling the history of evolutionary theory from Darwin to the end of the Modern Synthesis (Chaps. 2 and 3), which I place around the mid-1970s when Dawkins gave us the replicator-vehicle concept. This history draws upon the same narrative arc that many on both sides of the debate about the Modern Synthesis, commit to, but I also include a discussion of multi-level selection theory and inclusive fitness. I am aware that there are other histories to be told that have been told, and I think this is another way into the debates at the core of this book but one I am not skilled enough to produce.

After the history, I then deal with some core explanatory issues. First comes causation (Chap. 4) and in particular I focus upon Mayr's detailed working of the proximate–ultimate distinction and recent critics of this. I regard Mayr's classic paper on this topic as a central philosophical statement of the Modern Synthesis, and given the amount of critical effort directed toward it, my impression is that I am not alone. In Chap. 5 I introduce information in the context of plasticity, although this topic receives treatment throughout the remainder of the book.

Chapters 6, 7, and 8 then deal with arguments from those advocating for an Extended Evolutionary Synthesis. They are evolution and development, epigenetics and niche construction theory. In each chapter, I focus on some core positions in these fields, well aware that I cannot do justice to the volumes of fascinating material accumulated under each topic. But my hope is to pull out the kind of reasoning at work and to contrast that with my own interpretation drawn from my views on information, among other things. In Chap. 9, I bring all of this together with further discussion of information and also developmental systems theory, leading to a statement of my current position with regard to the issues at hand.

I have necessarily missed various arguments from my book, taking what might be seen as the developmental route through the extended synthetic claims. I do not discuss evolving genomes, macroevolution, the major transitions, nor issues emerging from structuralism. This does not indicate a lack of interest but rather a decision about how to make my case. I may return to these issues at a later date.

1.1.3 Conclusion

This book argues against an extended synthesis, but it does not represent the only way to do this. In making a negative case, I am not attempting to disparage the enormous efforts made by those advocating for the extension. To leave a theory uninspected is irresponsible. They are products of their time, and new discoveries may well challenge them, and at some point, make clear that change is required. Those working for extension are doing this and causing those of us who disagree to think long and hard about why we do. The Modern Synthesis is a framework theory, as I discuss in the final chapter, and is a product of work over an extended historical period. That makes its inspection and analysis difficult but rewarding work. My hope is that I engage the reader just as I have been engaged by all whom I cite.

References

Baedke J (2021) What's wrong with evolutionary causation? Acta Biotheor 69(1):79–89. https://doi.org/10.1007/s10441-020-09381-0

Baedke J, Fábregas-Tejeda A, Vergara-Silva F (2020) Does the extended evolutionary synthesis entail extended explanatory power? Biol Philos 35(1):1–22. https://doi.org/10.1007/s10539-020-9736-5

Bradbury RB, Stoate C (2000) The ecology of Yellowhammers Emberiza citrinella on lowland farmland. Ecol Conserv Lowland Farmland Birds:165–172

Crick F (1970) Central Dogma of molecular biology. Nature 227(5258):561–563. http://cs.brynmawr.edu/Courses/cs380/fall2012/CrickCentralDogma1970.pdf

Dawkins R (1989) The selfish gene, 2nd edn. Oxford University Press, Oxford

Delisle RG (ed) (2017) The Darwinian tradition in context. Springer Nature Switzerland. https://doi.org/10.1007/978-3-319-69123-7

Dickins TE (2020) Conflation and refutation: Book Review of Uller, T. and K. N. Laland. eds. 2019. Evolutionary Causation: Biological and Philosophical Reflections. MIT Press, Cambridge, MA. pp. ISBN: 978–0–262-03992-5. $60.00/£50.00. *Evolution*, 508–514. https://doi.org/10.1111/evo.13916

Dickins TE, Dickins BJA (2008) Mother Nature's tolerant ways: why non-genetic inheritance has nothing to do with evolution. New Ideas Psychol 26(1):41–54. https://doi.org/10.1016/j.newideapsych.2007.03.004

Dickins TE, Dickins BJA (2018) The extent of the modern synthesis: the foundational framework for evolutionary biology. In: Burggren W, Dubansky B (eds) Development and Environment. Springer, New York, pp 155–176

Eldredge N (1971) The allopatric model and phylogeny in paleozoic invertebrates. Evolution 25(1):156. https://doi.org/10.2307/2406508

Eldredge N, Gould SJ (1972) Punctuated equilibria: an alternative to phyletic gradualism. In: Schopf TJM (ed) Models in paleobiology. Freeman, Cooper and Company, New York, pp 83–115

Evans LT (1984) Darwin's use of the analogy between artificial and natural selection. J Hist Biol 17(1):113–140. https://doi.org/10.1007/BF00397504

Floridi L (2010) Information: a very short introduction. Oxford University Press, Oxford

Futuyma DJ (1986) Evolutionary biology. Sinauer Associates, Sunderland, MA

Gildenhuys P (2004) Darwin, Herschel, and the role of analogy in Darwin's origin. Stud Hist Philos Sci C Stud Hist Philos Biol Biomed Sci 35(4):593–611. https://doi.org/10.1016/j.shpsc.2004.09.002

Godfrey-Smith P (1996) Complexity and the function of mind in nature. Cambridge University Press, Cambridge

Gould SJ (1982) Darwinism and the expansion of evolutionary theory. Science 216(4544):380–387. https://doi.org/10.1126/science.7041256

Gould SJ (2002) The structure of evolutionary theory. The Belknap Press of Harvard University Press

Gould SJ, Eldredge N (1977) Punctuated equilibria: the tempo and mode of evolution reconsidered. Paleobiology 2(2):115–151. https://doi.org/10.1017/S0094837300005224

Gould SJ, Lewontin RC (1979) The spandrels of San Marco and the Panglossian paradigm: a critique of the adaptationist programme. Proc R Soc Lond Biol Sci 205(1161):581–598. https://doi.org/10.1098/rspb.1979.0086

Gould SJ, Vbra ES (1982) Exaptation—a missing word in the science of form. In Paleobiology (Vol. 8, pp. 4–15)
Grafen A (1984) Natural selection, kin selection and group selection. In: Krebs JR, Davies NB (eds) Behavioural ecology, 2nd edn. Blackwell Publishers, New York, pp 62–84
Huneman P, Walsh DM (eds) (2017) Challenging the modern synthesis: adaptation, development, and inheritance. Oxford University Press, Oxford
Huxley JS (2010) Evolution: the modern synthesis: the definitive edition. MIT Press, Cambridge
Lewens T (2019) The Extended Evolutionary Synthesis: what is the debate about, and what might success for the extenders look like? Biol J Linn Soc 127(4):707–721. https://doi.org/10.1093/biolinnean/blz064
Love AC (2017) Evo-devo and the structure(s) of evolutionary theory. In: Huneman P, Walsh DM (eds) Challenging the modern synthesis: adaptation, development, and inheritance. Oxford University Press, Oxford, pp 159–187
Mayr E (1982) The growth of biological thought: diversity, evolution and inheritance. Belknap Press
Mayr E (1989) Speciational evolution or punctuated equilibria. J Social Biol Syst 12(2–3):137–158. https://doi.org/10.1016/0140-1750(89)90041-9
Nettle D, Bateson M (2015) Adaptive developmental plasticity: what is it, how can we recognize it and when can it evolve? Proc R Soc B 282(1812), 20151005. https://doi.org/10.1098/rspb.2015.1005
Otsuka J (2019) Ontology, causality, and methodology of evolutionary research programs. In: Uller T, Laland KN (eds) Evolutionary causation: biological and philosophical reflections. MIT Press, New York, pp 247–264
Oyama, S. (2000) The ontogeny of information: developmental systems and evolution (2nd edn). Duke University Press
Pigliucci M (2007) Do we need an extended evolutionary synthesis? Evol Int J Org Evol 61 (12):2743–2749. https://doi.org/10.1111/j.1558-5646.2007.00246.x
Pigliucci M, Müller G (2010a) Elements of an extended evolutionary synthesis. In: Pigliucci M, Müller GB (eds) Evolution: the extended synthesis. MIT Press, New York, pp 3–17
Pigliucci M, Müller GB (2010b) Evolution: the extended synthesis. MIT Press, New York
Potochnik A (2020) Idealization and many aims. Philos Sci 87(5):933–943. https://doi.org/10.1086/710622
Ridley M (2004) Evolution, 3rd edn. Blackwells
Scott-Phillips TC, Laland KN, Shuker DM, Dickins TE, West SA, Scott-Phillips TC, Laland KN, Shuker DM, Dickins TE, West SA (2013) The niche construction perspective: a critical appraisal. Evolution 68(5):1–13. https://doi.org/10.1111/evo.12332
Smith TL (2018) The roles of the analogy with natural selection in B.F. Skinner's philosophy. Behav Process 161:139–148. https://doi.org/10.1016/j.beproc.2018.02.007
Smocovitis VB (1992) Unifying biology: the evolutionary synthesis and evolutionary biology. J Hist Biol 25(1):1–65. https://doi.org/10.1007/BF01947504
Smocovitis VB (1996) Unifying biology: the evolutionary synthesis and evolutionary biology. Princeton University Press
Stevenson A (ed) (2010) Oxford dictionary of english, 3rd edn. Oxford University Press, Oxford
Theunissen B (2012) Darwin and his pigeons. The Analogy between artificial and natural selection revisited. J Hist Biol 45(2):179–212. https://doi.org/10.1007/s10739-011-9310-8
Uller T, Helanterä H (2019) Niche construction and conceptual change in evolutionary biology. Br J Philos Sci 70(2):351–375. https://doi.org/10.1093/bjps/axx050
Uller T, Laland KN (eds) (2019a) Evolutionary causation: biological and philosophical reflections. MIT Press, London
Uller T, Laland KN (2019b) Evolutionary causation. In: Uller T, Laland KN (eds) Evolutionary causation: biological and philosophical reflections. MIT Press, London, pp 1–12
Vrba ES, Gould SJ (1986) The hierarchical expansion of sorting and selection: sorting and selection cannot be equated. Paleobiology 12(2):217–228. https://doi.org/10.1126/science.49.1254.50

Webb RH (2011) If evolution is the answer, what is the question? J Evol Psychol 9(2):91–107. https://doi.org/10.1556/JEP.9.2011.2.2
Williams GC (1996) Adaptation and natual selection: a critique of some current evolutionary thought, 3rd edn. Princeton University Press, Princeton
Wilson DS (1975) A theory of group selection. Proc Natl Acad Sci USA 72(1):143–146. https://doi.org/10.1073/pnas.72.1.143
Wynne-Edwards VC (1962) Animal dispersion in relation to social behaviour. Oliver and Boyd

Darwinian Evolution 2

Abstract

This chapter describes the emergence of modern evolutionary thought. It begins with Lamarck's transformational theory of evolution, a theory dependent upon core metaphysical assumptions. Darwin's triumph was the removal of those assumptions, through the pursuit of empirically derived scientific axiom and the introduction of population thinking. Nonetheless, he was unable to do more than black box the mechanisms of variation and inheritance essential to his theory of evolution by natural selection. This left the field open for differing positions on whether evolution was a gradual process, and whether variation was continuous. With the emergence of quantitative biology and statistical methods of analysis a long debate between Darwinian gradualists, committed to continuous variation, and saltationists who thought variation was discontinuous began. The rediscovery of Mendel's particulate model of inheritance in 1900 enhanced the saltationist position, even though there is some evidence Mendel saw continuous variation as underpinned by particulate inheritance. It was the resolution of this debate that has been regarded as the beginning of the Modern Synthesis in evolutionary biology, and so this chapter ends at this point. The Modern Synthesis is the focus of the following chapter.

Keywords

Evolution · Transmutation · Transformation · Inheritance · Variation · Differential success · Natural Selection · Gradual versus discontinuous variation · Biometricians · Mendel · Mendelians · Particulate inheritance

T. E. Dickins, *The Modern Synthesis*, Evolutionary Biology – New Perspectives on Its Development 4, https://doi.org/10.1007/978-3-030-86422-4_2

2.1 Introduction

This book is about theories of evolution. This means that evolution must be something that one can have theories about. Evolution must be recognizable to some considerable degree, but not self-evidently explicable. There is scientific and theoretical work to be done.

Evolution as we conceive of it today, emerged in Europe as a topic of philosophical and scientific enquiry during the eighteenth century. That emergence was from a culture of Christian belief in which God created the diversity of life that could be seen on the planet. Studying that diversity was a respectable pursuit, and common among the clergy, but initially each form of life encountered was seen as a perfection, created by God. This idea was perhaps most famously captured by Paley's argument from design in his 1802 book, *Natural Theology or Evidences of the Existence and Attributes of the Deity* in which he made comparison between the intricacy of contrivances designed by humans and the intricacies of biological forms, such as the eye. A watch is clearly performing a purpose and does so through mechanisms invented by a designer. Similarly, the eye is mechanically structured to perform a purpose, entailing a designer and that designer could only be God. This belief was not held without challenges, and there were different versions of God's involvement from one of total control over all form through to the creation of natural laws that were then left to run their course. It is into this latter view of divine causation that modern views of evolution were inserted, and then developed.

The term *evolution* colloquially means change over time, a form of developmental process. While the process of development from infancy to adult undoubtedly did not go unnoticed, the view of evolution as the development of different types of life over historical time is surprisingly recent. Mayr (1982) credits Lamarck, emerging from the end of the eighteenth century and into the 19th, for "converting the unbroken chain of a created sequence of ever greater perfection into a line of descent" (p.343). Lamarck focused upon two issues: the gradual increase in perfection (complexity) from the simplest to the most complex organisms; and the diversity of organisms. Both of these phenomena were there to be seen by all, what Lamarck added was the idea that one species could, over very long time, transform into another. Thus, there were marked differences between organisms, but something of the simpler could be seen in the more complex suggesting transition. The gradual nature of this evolution was such that change was imperceptible in real time and it was only later, further down the phyletic line, that variation becomes apparent.

According to Mayr, Lamarck developed these ideas as a result of studying the mollusk collection in the Paris Museum, which he inherited as curator in the 1790s. By organizing specimens through geological strata, and thus time, he was able to discern the slow phyletic change, and to make comment that this resolved a pressing issue at the time, that of extinctions. The scientific community recognized that different organisms had inhabited the Earth in the past, but some had trouble aligning their loss with an omnipotent God. Lamarck argued that the introduction of long, historical time enabled change, and that change could be quite radical. Furthermore, it enabled the God of the natural theologians to have designed organisms that could adapt to a changing world. Whether this meant that Lamarck thought God myopic and unable to predict all future change in His creation is unclear, but this view rendered extinction a non-problem. Species had not been removed, merely adapted to new circumstance.

Lamarck's gradual *transmutation* across long historical time not only provided adaptation to a changing world, but also led to the emergence of different kinds of organism, transforming along separate branches. This concept was in radical opposition to the Great and unbroken Chain of Being that most natural theologians held as a truth. Contemporary flora and fauna were not representatives of constant species, unchanging across time. Nonetheless, new branches were not spontaneously formed, and the only spontaneous emergence of new forms was hypothesized early in the emergence of life. Thus, the simplest of organisms contained all that was required to enable continued change and adaptation.

How were Lamarck's evolutionary changes caused? First, Lamarck assumed that life had the intrinsic power to acquire complexity, or perfection, and this was something endowed by God. The second cause is one that perhaps attracts the most attention from contemporary scholars. Mayr summarizes it as follows:

> The need to respond to special circumstances in the environment will... release the following chain of events: (1) Every considerable and continuing change in circumstances of any race of animals brings about a real change in their needs ("besoins"); (2) every change in the needs of animals necessitates an adjustment in their behavior (different actions) to satisfy the new needs and, consequently, different habits; (3) every new need, necessitating new actions to satisfy it, requires of the animal that it either use certain parts more frequently than it did before, thereby considerably developing and enlarging them, or use new parts which their needs have imperceptibly developed in them "by virtue of the operations of their own inner sense" (Mayr 1982: 354).

This second cause is behavioral, and it is responsive to the demands of the environment. There is no hidden, long-term objective but merely the requirement to maintain harmony between the organism and the world, as such it is a homeostasis concept. Lamarck's First Law of Use and Disuse allows those animals that still retain developmental capacity to increase the complexity of their design through behavioral response. This responding is mechanistic—there is no sense of vitalism at play in this idea. His Second Law of the Inheritance of Acquired Characteristics enables newly developed complexity, new adaptations[1] to be passed onto the next generation just so long as those changes are shared by both sexes or displayed by the sex that produces offspring.

By Mayr's account none of Lamarck's ideas are original to him, but instead were present in many works of natural philosophy prior to and during his time. This includes the ideas enshrined in both of his Laws. They become lawlike under Lamarck because he recruits them to explain gradual evolutionary change toward

[1]It should be noted that this use of "adaptation" is not a formal one. Adaptation to the changing world is the focus of Lamarck's theory, not the creation of adaptations. Thus adaptation was the final result of his causes working in order to maintain harmony between organism and world.

perfection, and to solve problems like extinction. In many ways, Lamarck's role as a curator of mollusks was extended to a careful curation of ideas. This introduced weakness to the framework. He adopted Cartesian dynamics to explain the production of behaviors required by his first Law, which necessarily proved false. Furthermore, he felt no need to provide a mechanical account of the inheritance of acquired characteristics, as it was regarded as self-evidently true at the time: offspring resemble their parents. What he provided was a reason why characteristics were acquired and that was to maintain harmony in a changing world.

Mayr's account of Lamarck summarized here, places the emergence of *evolution* as a fact squarely at this point in history.[2] Lamarck comprehensively moved away from the idea of a static world, of essences fixed by God, to one of change. Organisms had to keep up with environmental change and could do so through mechanical response that led to developmental adjustments, which in turn were inherited. Over time the accumulation of developmentally induced change displayed itself as transition in the *form* of organisms. We are thus left with a view of what evolution is and how it occurs.

Of course, Lamarck's efforts were not universally accepted at the time, or for many years thereafter. The dominant natural theology of the eighteenth century still had a grip on the scientific imagination, and it took further developments in natural history, including improving fossil records, for evolution to gain acceptance among scholars. Once evolution became a topic of scientific discussion and enquiry some decades further into the nineteenth century, then different theories of evolution emerged (Delisle 2019; Desmond and Moore 1991; Mayr 1982, 1991).

2.2 Darwin's Contribution

Darwin's scientific efforts were firmly grounded in natural historical methods, an approach he adopted during childhood and pursued thereafter with great diligence, including wide reading in natural history, natural theology, geology, and also the adventures of others, including Humboldt (Mayr 1982). His interest was in the diversity of life and he came to place that interest within an evolutionary context. His engagement with evolution was doubtless due to many exposures that he had to contemporary debates while a student both at Edinburgh and Cambridge universities (Desmond and Moore 1991) but also due to interactions with leading scientists when making sense of his *Beagle* collections after 1836 (Mayr 1991).

It was his voyage on the *HMS Beagle* (1831–1836) as a gentleman companion to Captain Fitzroy, and later as the adopted ship's naturalist that perhaps had the greatest effect (although see Van Wyhe 2013) for an alternative view on Darwin's appointment). Whether that effect was initiated during the voyage or entirely afterward is a point of historical discussion (Brinkman 2010; Sulloway 1982a).

[2]Lamarck's main book (Philosophie zoologique, ou exposition des considérations relatives à l'histoire naturelle des animaux) was published in 1809 in Paris, the year of Darwin's birth.

But, by 1837 historians agree that Darwin had entirely shifted from previously held, orthodox views on the fixity of species to a version of transmutation of species.

Brinkman has Darwin pursuing a research agenda on his voyage set by Lyell's *Principles of Geology*, a book that promoted an essentialist and non-evolutionary view of nature. Lyell argued for a cyclical process of climatic change, leading to extinction of those animals less able to cope with change, and then the introduction of novel species to fit gaps left by extinction. There was no gradual increase in perfection; there was no adaptation even though new introductions were better fit to new climactic conditions; but, there was a uniformity of process in this view and it led to the agenda Brinkman referenced (Mayr 1982). This permitted researchers to focus upon the causes of extinctions and subsequent introductions, and this in turn led to a particular ecological focus. Lyell himself offered no stated cause of new introductions, leaving room for creationist views. While he had evidence of extinctions, there was no observational record of introductions, and Lyell's philosophy of science demanded such evidence. He could only infer it was a regular process.

Darwin will have collected his numerous specimens, executed his many dissections, and written in his notebooks with Lyell's theory firmly in mind, this much is agreed. He was also well versed in the natural theology of Paley, and the idea of God as a designer. Darwin's notebooks and other sources have received forensic examination in order to determine when Darwin began to have doubts about Lyell (post-*Beagle*: (Sulloway 1982a) or during (Brinkman 2010)) but whatever the truth of it, a certain kind of question at some point occurred to Darwin, after exposure to much data and many ideas. Desmond and Moore (1991) capture this by siding with Sulloway and suggesting Darwin suffered a revelation while listening to Lyell address the Geological Society in London. Lyell was discussing Owen's view that fossil creatures are closely related to modern animals. Accordingly, Darwin realized that fossils he had found in South America were not of European or African species but were instead specific to that area. These creatures were similar, but crucially those found in South America were different and most closely related. How could this happen?

Darwin's question is about similarity, and thus diversity, but it is also about geography (and ecology in modern parlance). He developed vertebrate fossil successions and became fixated on this question of geographical specificity and relatedness across time. As Brinkman notes, this could be answered by Lyell's cycles, but Darwin also knew of Lamarckian and other transmutational ideas, and he explored those possibilities. In essence, Darwin became intrigued by the possible causes of change. Moreover, he wondered how the observed diversity of extant life might be related to these views.

Darwin's views on extant diversity are often captured by discussion of his extended exploration of the Galapagos archipelago during his *Beagle* adventure. The most common historical trope focuses upon Darwin's observations of finches, suggesting that he became aware of diversity across the archipelago, and specific adapted variation on particular islands.[3] Thus, finch species had beak morphologies that suited them to food extraction problems endemic to particular ecologies. Here again we see the role of geographical distribution of variation. This is an observation that Lyell could resolve in terms of new introductions meeting local conditions, but in that answer there is no mechanism of origin. In keeping with the late conversion hypothesis, Sulloway has questioned this trope claiming that Darwin did not fully master the variation present in his finch samples until he was returned to London and in the presence of Gould, a notable ornithologist (Sulloway 1982b, 1984).[4] Gould made clear that the finches were of different species on each of the different islands of the Galapagos, and that none of these species were found on the mainland of South America; they had not simply come across on the trade winds in the forms Darwin had found. As a result, Darwin came to see the finches as a useful example of the core problem—that of diversity and its cause.

Darwin was with Gould, and other experts from 1836, and by 1837 he was actively pursuing evolutionary possibilities for his problem. And then, in 1838 he read Malthus' *An Essay on the Principles of Population.* In Darwin's own words:

> Fifteen months after I had begun my systematic enquiry, I happened to read for amusement Malthus on Population, and being well prepared to appreciate the struggle for existence which everywhere goes on, from long-continued observation of the habits of animals and plants, it at once struck me that under these circumstances favorable variations would tend to be preserved, and unfavorable ones to be destroyed. The result of this would be the formation of new species. Here, then, I had at last got to a theory by which to work (Quoted in Mayr 1991: 70).

That theory was to be named *evolution by natural selection.*

Malthus had noted that human populations tend to geometric, or exponential growth due to their reproductive habits, while resources at best increase in a merely linear fashion. At some point population outstrips the carrying capacity of the resources and this leads to conflict, famine, disease, and death. In this way the population is checked by the extrinsic, environmental reality of resource, with only some individuals having the capabilities to weather the struggle for existence.

Within the historical literature there is much debate about just how much of an epiphany the reading of Malthus caused within Darwin (Mayr 1991; Ruse 1975; Schwartz 1974). Whether one believes Malthus revolutionized Darwin's thinking, or that he was prepared by much intellectual development to make best use of Malthus, it is clear that Malthusian ideas were central to the development of Darwin's

[3]It is most probable that differences between species across the islands were raised with Darwin by the Vice Governor of the archipelago. Darwin himself noted differences in mockingbirds collected from two islands (Sulloway 1982b).

[4]This is possibly due to the fact that Darwin failed to keep track of the location of his collected species on the Galapagos, and this information was pieced together after the voyage from the more assiduous records of Captain Fitzroy. An odd lapse in Darwin's otherwise meticulous habits (Reznick 2010).

explanation. Put simply, Malthus enabled Darwin to fully develop and introduce population level thinking to his views on transmutation, or evolution.

Populations were to be understood as collections of individuals, within a species, that varied relative to one another; something Darwin knew from his life in the country as well as his interactions with numerous animal breeders and taxonomists. Now those individuals were in direct competition with one another for crucial resource—an idea not previously entertained by naturalists who had rather thought of whole species as in competition—and some individuals were finding their specific niche as a result. Individual variation could cause differential success in that competition, and those variants that thrived would be inherited and passed along, as animal breeders had shown. For the breeders this was artifice, they selected what to favor. Darwin saw that the economics of nature led to the overrepresentation of favored types, a process that looked like selection. Over time the characteristics of the population would change as a result of this differential success.

2.2.1 Evolution by Natural Selection

Darwin's contribution is a statistical one, effectively noting that individuals have different probabilities of survival within a competitive population. Where Lamarck's First Law has organisms actively seeking homeostasis through behavioral response, Darwin has a population full of individuals that vary along trait dimensions. Lamarck's creatures grapple with a changing physical environment; Darwin's environment is all encompassing and includes competition with other individuals. Lamarck's creatures actively solve the problems of the physical environment, whereas Darwin's either do or do not have the capacity to thrive as a consequence of their naturally bequeathed characteristics. Both Lamarck and Darwin rely upon a principle of inheritance to transmit newly acquired or relatively successful characteristics to the next generation, thereby permitting evolutionary change.

Mayr (1991) analyzed Darwin's explanatory model, from which natural selection was derived, in order to yield an abstraction. He noted that the model consisted of five facts (F) and three inferences (I). The first three facts (and their sources) were:

- (F1) the possibility of geometrical increase in populations (Malthus).
- (F2) the stability of populations (observations from multiple sources).
- (F3) resource limitation (Malthus).

The first inference:

- (I1) that there is a struggle for existence between individuals (Malthus).

And then, two more facts:

- (F4) the unique nature of individuals (animal breeders and taxonomists).
- (F5) the heritability of individual variation (animal breeders).

Leading to Darwin's own contribution:

- (I2) Differential survival, the core of natural selection.
- (I3) Evolution across many generations.

Huxley presented a similar derivation (Huxley 2010):

- (F1) tendency of all organisms to increase in geometric ratio.
- (F2) nonetheless, numbers of a given species tend to remain constant.
  Leading to the first deduction (D):
- (D1) there is a struggle for existence.
  Then:
- (F3) all organisms vary appreciably.
- (D2) natural selection (from F3 and D1).
- (F4) some variation is inherited.
- (D3) the result of the above is the differential transmission of inherited variation.

Huxley notes that two forms of natural selection can be isolated from his version of the model, and the same is true of Mayr's. The first kind is that of natural selection in the broad sense, which is purely the idea of differential survival. The second, natural selection in the evolutionary sense, becomes the differential transmission of inheritable variations across generations. Both forms of natural selection are purely a consequence of how variation meets the struggle for existence. When inheritance is added it is clear that evolutionary change is a consequence of a blind mechanism, quite different to Lamarck's agent-led view of change.

Lamarck's view of evolution, emerging from the eighteenth century and the natural theological tradition, rests on certain assumptions. Those assumptions are to some extent metaphysical in flavor, where we understand metaphysics in a scholastic sense as principles that are necessarily true (van Inwagen and Sullivan 2020). Thus, the inheritance of acquired characteristics and the need to maintain harmony in the universe were regarded as self-evident, as was the existence of a deity that was ultimately responsible for all. Given these constraints it is perhaps not surprising that Lamarck relied upon the designed agency of organisms to solve the problems of environmental change. Lamarck's God, rather than being myopic, may have recognized the emergent properties of a dynamic Earth, and matched them with flexible life. From this bedrock Lamarck sought to organize principles and data to account for evolution.

Darwin did not rely upon scholastic metaphysics to ground his thinking but was instead focused upon collecting data and thus evidence from which to draw conclusions. In this way Darwin's program of work was staunchly empiricist; the data would lead to a point where the inferences abstracted by Huxley and Mayr would be justified. Ruse, writing in an introduction (Reznick 2010), characterizes Darwin as an inductivist dealing with the ideas of Herschel and Whewell, both among his mentors. For these two philosophers the job of science was to derive axiomatic laws and to find the true cause (*vera causa*) of a thing. They venerated Newton for his deductive use of axioms but also for gravity, a cause that explained multiple phenomena in the natural world. Herschel understood the *vera causa* of gravity as something directly observable when objects were dropped, etc., but he also believed that gravity was created by God's will. Whewell did not require direct observation to derive a "consilience of inductions" that would tie together diverse phenomena and argued that indirect evidence was sufficient to build a science of true causes.

Ruse interprets Darwin as more Herschellian. His reliance upon the artifice of animal breeders, to select new forms according to whatever preference, was an attempt to directly observe selection. While artificial selection might be considered an analogy for natural selection, it was perhaps only so because no agent was required in the latter case. All else was equal. Thus, Darwin was eager to remove all metaphysics as a foundation to his ideas. Darwin's instinct was to self-contained, contingent, and empirical truths. He was trying to produce his own axioms and laws and natural selection should be interpreted in this light.

There were two self-evident truths that Darwin did rely upon throughout the development of his theory and the *Origin*: that of variation and heredity. Variation was measurable, but its source was unknown. The resemblance between a parent and its offspring was sufficient to justify heredity, but the causes of this too were unknown.

2.2.2 Variation and Heredity

Variation and heredity are concepts that are potentially in agonistic interaction with one another (Mayr 1982). For offspring to resemble their parents then parental traits must be passed on such that those traits are similar to some high degree across the generations. This implies that heredity is what Mayr refers to as *hard*, resistant to change, and in a state of at least semi-permanence. For Darwin, when he developed natural selection, it was those traits that enabled organisms to thrive under a Malthusian struggle for existence that would then be passed on to the next generation. Gradually, generation after generation, more and more of those traits would be represented in the population. Hard inheritance was required.

For evolutionary change to occur over time, as circumstances changed in the environment, for example, then new variation was required to provide new solutions to new problems. That which was inherited in these circumstances, was different to that which had been inherited previously. Where did this variation come from? Was some inheritance at least malleable, or *soft* like Lamarck's, where the agent can behave and induce new variation and then pass it on to the next generation?

> It is frequently stated that Darwin totally rejected all Lamarckian ideas in the first edition of the Origin and that he allowed at that time no mechanisms of evolution other than random variation and natural selection. This is not correct. Darwin displayed considerable indecision already in 1859 on the origin of variation and the nature of inheritance. He makes no less than three sets of concessions to the possibility that the environment in the widest sense of the word can induce genetic variation and that these acquired characters can be inherited. First, he speculated about a direct effect of the environment on certain structures; second, he

> hypothesized an indirect effect of the environment in increasing variability; and third, he discussed the effects of use and disuse, for instance, when he says that the reduced size of the eyes of moles and other burrowing mammals is "probably due to gradual reduction from disuse but aided perhaps by natural selection" (Mayr 1991: 109).

At the same time Darwin was quick to rule out Lamarckian cause for certain adaptations found in the social insects, according to Mayr.

While Darwin believed in both hard and soft inheritance, and also that there was most likely a large reserve of available variation in populations but with the odd, infrequent emergence of new variation, he never solved these problems. He relegated the issues of variation and inheritance to a black box during the development of the *Origin* (Mayr 1982). These things were essential for his model of evolution, they must be happening, and he knew new variants would pop up during the artificial selection operated by breeders indicating that novelties could emerge but nonetheless the hard aspects of inheritance were sufficient for those breeders to meet their goals. This was not quite the Herschellian derivation Darwin might have wanted, but it was sufficient in Whewellian terms as indirect evidence. Darwin returned to these issues in 1868 when he published *Variation of Animals and Plants under Domestication.*

In *Variation* Darwin proposes *pangenesis*, a collection of ideas designed to deal with numerous problems that had accumulated since the *Origin*, including variation and inheritance. Pangenesis was an attempt to derive a *vera causa* to tie them all together.

There are two major components of the pangenesis package (Mayr 1982):

1. *Gemmules*—very small particles that cause the development of traits and convey their heritable qualities. Each cell has an idiosyncratic gemmule and parental contributions of gemmules can mix, allowing for hybridization and also the emergence of previously dormant types. In effect, the relation of gemmules to one another can be one of dominance or recessiveness, such that certain traits are expressed in preference to others under specific conditions.
2. *Transportation hypothesis*—this is to account for the effects of use and disuse. Gemmules will circulate throughout the system and will self-replicate when suitably nourished and develop into the kinds of cells from which they originated. These gemmules can accumulate in the reproductive system, to be passed on. And, crucially, these gemmules can be modified by new conditions and those modifications can be passed on by reproduction, should they accumulate in the reproductive system.

As can be seen, the first component is one of hard inheritance, and the second incorporates soft inheritance. The environmental modification of the gemmules, seen in component 2, might be conceived of as a model of environmentally induced mutation, but the modification appears directed to achieve fit with the environment rather than being truly random and relying upon selection alone. It is the directed nature of malleable change that Mayr captures with the term soft inheritance, and as such it conforms to Lamarckian ideas.

2.3 The Darwinian View of Evolution

In Sect. 2.1 we saw that Lamarck's interest was in the diversity of forms found in nature. Moreover, he recognized a transition of increasing complexity across forms that he was able to organize temporally due to where his mollusks resided in the geological strata. For Lamarck and other transmutationists, form changed across time and what was required was a theory for how that change occurred. Change was achieved by the agency of the organisms themselves, and those behavioral adaptations that proved useful were passed on to offspring and subsequent generations.

Darwin too was interested in the diversity of form and came to realize that a transmutational approach would help to explain facts such as the different species of finch on the Galapagos, with no known representatives on the mainland. But in order to derive an account Darwin began to think of populations as groups of individuals, that varied in ways that might assist them in the struggle for life. Thus, natural variation supplied the opportunity for natural selection to occur. Populations became populated with particular types, particular variants, because of their success in this Malthusian struggle. Those variations were heritable and inherited. Here is the key difference. While Lamarck understood that perfected forms must be inherited for them to remain in the population, he did not see the population as that which evolved. Darwin did.

Evolutionary theory is thus focused upon the perpetuity of certain forms across time, and the introduction of new ones. But where Lamarck saw this as resulting from the efforts of individuals to solve problems, he did not incorporate the dynamics of relative success between individuals. Darwin's population thinking was truly revolutionary, but it was also embedded within a contemporary philosophy of science that sought empirical support and to derive parsimonious causal constructs from that evidence, rather than to rely upon metaphysical constructs as a foundation for explanations of observations.

Darwin's problems with variation and inheritance, and his attempt to resolve them with pangenesis, demonstrate that form was central to his theory. But his theory was not about the generation of form. For natural selection to work there must be processes that can develop new variants in form and pass those new variants on. His instinct was to produce mechanistic hypotheses, in keeping with the views of his philosophical mentors. But his black boxing of those processes while he developed the *Origin* tells us something useful about how his theory of evolution worked: it simply required variant forms in the context of Malthusian struggle to enable natural selection. Selection would then sculpt populations with particular characteristics. Population structure was given explanatory priority because for Darwin variation only made sense within the context of a population.

At this point in theoretical development the fact that there was variation and inheritance was sufficient to make the case for natural selection. The activities of breeders and their artificial selection procedures were an existence proof for selection once the preferences of the agents (breeders) were replaced with the contingent problems of existence. This conceptual framework in many ways afforded much latitude to any theories of variation and inheritance; just so long as they explained those phenomena they could be fitted to Darwinian evolution. But the expectation was that any theories would be empirically grounded.

Perhaps less obviously the black boxing, or bracketing, of variation and inheritance showed that while they were logically related to evolution by natural selection as phenomena, their causal explanations were not themselves relevant. Put another way, multiple theories of variation and inheritance could be falsified but just so long as we were sure of the phenomena then this would not impact upon the Darwinian theory of evolution. This is not an available route for falsifying this theory. This makes clear why Darwin was able to incorporate some elements of Lamarckian soft inheritance into his speculations, and it suggests a reason why he was never quite able to deliver a full theory of variation and inheritance: his own theory did not provide the resources to generate one. That would require further empirical work, which Darwin did not pursue.

This does not mean that the concepts of variation and inheritance are entirely neutral with relation to the fate of the theory of evolution by natural selection. Variation has to be able to facilitate some form of advantage in biotic competition between individuals, and also in dealing with abiotic challenges from the environment. Moreover, that variation has to be inherited in a manner that enables advantageous traits to have traction down the generations and directly link to survival (Darwin's main emphasis) and reproduction. At the time Darwin was working this was an open field, and thus a large conceptual space, but nonetheless one with parameters (see Sect. 2.5 for a discussion of the biometricians and their interpretation of these parameters).

Given this analysis I believe it is possible to argue that what Darwin did was to fully establish the phenomenon of evolution, evolution as something that happens to populations, in a way far more definitive than Lamarck's assertion of transmutation as a fact. Lamarck noted mere similarity between organisms, and a temporal sequence of change. This was essential work to permit thoughts of transmutation, but his approach was to develop a theory of how variation is produced, through agency, and to rest that upon the metaphysical assumptions associated with inheritance as resemblance. To this end, Darwin's theory of evolution is a revelation about an almost lawlike process, that of natural selection.

2.3.1 The Challenge of Darwinian Evolution

Mayr (1982) describes Darwin's contribution not as a revelation but as a revolution. This is because the Darwinian scheme had broad philosophical implications. By siding with those who saw the world as a changing and dynamic place, rather than a static one, Darwin also removed teleological explanations of form and its fit with the world: selection sifted the winners and the losers in an unguided, statistical process and variation was always available to resolve new contingent problems of existence. Darwin saw selection as enabling gradual descent with modification across many, many generations and into this he included humans as primates descended from ape-like ancestors rather than as created in the image of God.[5] And Darwin broke with the Aristotelian essentialism that had gripped natural philosophy for so long, by introducing population thinking.

Mayr also lists Darwin's philosophy of science, his hypothetico-deductive method developed in the context of Herschel and Whewell, claiming Darwin as the first scientist to deal with this topic. For Mayr this is important in analyzing the reception of Darwin's theory. Where previously philosophers and the philosophically oriented, including Lamarck and his metaphysically grounded work, had kept speculations about transmutation within conceptual bounds Darwin relentlessly pursued data and grounded his theory building in the real, tangible world. Mayr claims this forced anyone willing to think about the diversity of life into making a choice between three possible explanations: a continual creation in keeping with Lyell's views on extinction and replacement; evolutionary laws instantiated by God; or Darwin's theory.

As Ruse recounts in his introduction to Reznick's treatment of the *Origin* (Reznick 2010), the *Origin* was met with staunch criticism as well as staunch advocacy in the years immediately following its publication. But the criticism was due to the role of natural selection, not the idea of evolution *per se*. For much of Victorian society, the default belief was one of the first two options given by Mayr: some form of creationism. Many regarded the more general idea of descent with modification—the slow transitions in form across time—as acceptable, as it is something that God could organize and they had prior exposure to various transmutational ideas (Provine 2001). Darwin's own advocacy of natural selection, against a backdrop of mid-Victorian creationism, involved relentlessly producing examples of organisms whose presence at precise points on the Earth could readily be explained by natural selection but less so by the interventions of an all-seeing and all-powerful God. For example, Mayr (1982: 504) quotes Darwin asking why "has the supposed creative force produced bats and no other mammals on remote islands?"

The acceptance of evolution after 1859 is perhaps best understood as acceptance of the kind of transmutation that Lamarck focused upon. That is, formal (or morphological) change over time was accepted as an empirical fact, due to fossil evidence and the kind of data collected by Darwin. But the *vera causa* of those forms was God, either directly or through divinely set process. This level of empirical acceptance is perhaps due, as Ruse notes in his introduction, to a gradual separation

[5]Thomas Huxley was the first to do this in his 1863 *Man's Place in Nature*. Darwin hinted at the human story in the *Origin*, but it was not until the *Descent of Man* (1871) and *On the Expression of Emotions* (1872) that he laid bare his full view.

of the spiritual and physical worlds in terms of theological dominion. The latter was the province of science, the former not. What Darwin did, for his critics, was to overextend the jurisdiction of data led science into matters still regarded as metaphysical and supernatural.

2.4 On Definition and Natural Selection

In Sect. 2.2.1 two abstractions of Darwin's derivation of natural selection were presented. Both abstractions make clear that natural selection is a consequence of reproductive habits, a struggle for existence, and available heritable variation. The differential success we see across all individuals is due to trait variation, and because traits are heritable those that are differentially successful will continue across the generations until circumstances change. This impacts upon the nature of population—what kinds of traits are represented within it. And, over long historical time, and the gradual accumulation of new variants, the types of individuals seen in the present differ radically from those seen in the past.

What both abstractions do not do is to simply locate natural selection at the point of selection, or elimination of less successful individuals. Selection, as a quotidian term, can be seen as sifting but natural selection is the production of variation and sifting within a population.

The Aristotelian approach within natural philosophy and science has been dominated by a form of essentialism (Popper 1945) in which the basic task was seen as creating a table, or record of all natural kinds. In this way knowledge was seen as an accumulation of ontological categories, and each natural kind required a definition that captured its fundamental essence. Popper discussed how this view has pervaded our thinking, even after the scientific revolution in which essentialism was directly attacked, and we find it in our common practices of definition.[6] Thus, the default response is to define *X* in terms of a list of properties, *1* to *n*. To that end we read our definitions from left to right. The idea of *X* is not questioned once a definition is requested. However, Popper argues that we should really process our definitions from right to left, and ask instead what we might call a thing that has properties *1* to *n*. In this way we empirically ground our ontology in measurable properties, and we are not seeking properties for an assumed natural kind. So, we do not ask, "what is natural selection?" but rather, "what should we term the outcome of geometric reproduction, the struggle for existence and the production of heritable variation that may assist in that struggle?" We need a term to aid exposition, but we do not begin with the term. This is his nominalist view of definitions.

[6]For Popper the problem with essentialist definition was that it implied access to truth and empirical certainty, something he could not commit to due to his views on what scientific knowledge was (Büttemeyer 2005). As Büttemeyer details there are some confusions in Popper's complete account of essentialist and nominalist definitions. I adopt Popper here to force the point that a definition is a term of art, a working procedure to get science done, and a term that can be changed by the doing of that science; it is not a statement of final truth.

Darwin discovered the process of natural selection through hypothetico-deductive science, through painstaking empiricism that led to him seeing how the world was and proposing a lawlike, statistical interaction. Each component of natural selection was itself a natural kind—variation and inheritance were measurable, and the nature of reproductive growth and population stability were also. All of these properties were lined up, in a relation to one another, and a term was applied to the derived process. This is what Huxley and Mayr showed us (Sect. 2.2.1).

To fully grasp the meaning of the term *natural selection* one has to read the right-hand side of its definition. But as Popper noted the term, as all scientific terms, is really only a form of scientific shorthand. The creationists who did not like natural selection perhaps fully understood this and saw that how it was defined left no room for God. However, it is possible to adopt the term without this effort and instead to use colloquial associations to discern its role in theory. Thus, for example, Lyell could only see natural selection as an eliminative force, sifting things out, rather than as a creative force that combined the production of new variation with differential success and reproduction down the line (Mayr 1982). Lyell's error here was probably compounded by his belief in replacement events, following extinction and thus a brand of creationism that saw God as the creator of novelty. Thus, he saw form (morphology) as an immutable, essential natural kind.

2.5 Variation and Biometrics

While Darwin did not need to fully explain variation, but merely to assume it (Sect. 2.3), he was convinced that individual variation was essential for his form of gradual, incremental evolutionary change (Joshi 2017). Two of Darwin's allies took issue with this commitment, Thomas Huxley and Galton (Darwin's cousin). In their view evolution occurred due to discontinuous leaps (saltation) in form, an idea that fed into the development of mutation theory by de Vries (Allen 1969).

Huxley's argument was simply that saltation explained the gaps in the fossil record, whereas a commitment to gradualism could not readily account for this, and he informed Darwin that his model of continuous variation added unneeded complexity to his theory (Provine 2001). Huxley emphasized his view in correspondence with Lyell, pointing out that novel breeds, such as the Ancon sheep had appeared in full form and at once, following the relevant parental cross. This use of artificial selection, the strong analogy that Darwin himself used to promote natural selection, is interesting. Again, it demonstrates that the bracketing of how variation occurs, during the development of the theory of natural selection, could permit other views on variation without upsetting the overall theory. Huxley accepted natural selection, but not the pace of change that Darwin argued for.

Galton's principal contribution to the variation issue was a statistical one.[7] He collected much data on traits such as human height and the size of sweet pea plants.

[7] Galton's main work was *Natural Inheritance* (1889).

What he noted was that populations over time appear to regress or revert to the median value of any given trait.[8] So, height would be a normally distributed trait within a population, with most individuals grouped toward the center, and fewer towards the extremes of short and tall statures. Galton observed that no offspring from such a population would produce a height outside of this range and that, on average, offspring would tend to be grouped toward the middle even if their own parents fell nearer to the extremities. Thus, the child of tall parents was most likely to be closer to the average height of the population and less likely to be tall or short.

This idea of regression, or reversion, was in fact first used against Darwin by Jenkin in 1876 (Joshi 2017). Jenkin argued that species had a fixed range of variation, and that variation was produced at random. As a result, the likely next variant would be somewhere near the middle of a normal distribution, rather than at the extremes. This idea was predicated upon a notion of fixed species type, and was therefore fundamentally in keeping with traditional, essentialist views. Nonetheless, Jenkin made other prescient observations about the loss of variation under blending inheritance and also from sampling error arising from a small number of individuals surviving relative to the almost infinite set of possible gametic combinations. As Joshi suggested, this is close to modern arguments about genetic drift. What Jenkin did was to clarify the parameters within which Darwinian selection would work.

As with Jenkin, Galton's regression was thought to be enough to undermine any work that natural selection might achieve with continuous, gradual variation. While a particular value might do better in one generation, the benefit of that trait would not be conferred to any offspring on average as they tended toward the center of the distribution. Thus, for natural selection to gain purchase significant discontinuities of variation would be required, something Galton and others referred to as *sports*, a term adopted by breeders.

For Galton, *sports* were the outcome of radical change in the stability of morphological equilibrium. They were to be contrasted with what he termed "variation proper" (Provine 2001), which simply captured some kind of strain upon normal conditions. These strains were insufficiently forceful to shift equilibrium points in morphological space but might perhaps be interpreted as wobbles about a stable point. Selection could operate on variation proper within a generation, thus enabling some variants to reproduce, but that reproduction would cause regression and would not impact upon long-term, multi-generational evolution.

Galton's statistical approach to evolutionary questions inspired two entirely separate camps: the biometricians, who wholeheartedly endorsed his quantification of morphological traits but also sided with Darwin's view of evolution as gradual and due to continuous, small variation; and the Mendelians who were also focused upon quantification but saw Huxley and Galton as correct with regard to discontinuous variation, or saltation. This divide emerged from a lack of clarity in Darwin's own writings:

[8] Galton used the term median, but more commonly people refer to regression to the mean. As this is about normal distributions both measures of central tendency should be equal.

> Thus, although Darwin's notion of descent with modification was widely accepted very rapidly after the publication of the Origin, his mechanism of natural selection remained under a huge shadow of doubt for many decades. Basically, there was an ambiguity in Darwin's formulations that could not be clarified in the absence of any knowledge of the mechanisms of inheritance. The ambiguity was about whether selection acted through a sort of a diffusion process in which alternative (discrete) trait variants replaced one another over time, or whether selection somehow directly modified the frequency distribution of a continuously varying quantitative trait. While Darwin favored the latter view, some passages in his writings lean toward the former, too. This ambiguity remained the basis for the bitter biometrician-Mendelian debates of the early 20th century (Joshi 2017: 531).

The debate between the biometricians (or orthodox Darwinians) and Mendelians began before the rediscovery of Mendel, and was, in effect, a debate between those invested in continuous or discontinuous variation prior to the rise of Mendelian concerns. Those focused upon discontinuity were at times labelled saltationists, as evolution was seen as progression *per saltum*, or in jumps. Both sides were focused upon the species question.

> The orthodox Darwinists believed that new species originate through the accumulation of small and continuous variations by natural selection. The saltationists, also called "mutationists," believed that new elementary species were created by sudden changes in heredity. This latter view implied that natural selection did not itself create new forms of organisms but merely selected between existing differences which were created by other causes like hybridization and mutation (Roll-Hansen 1989: 306).

The full list of contentious questions across the whole debate has been listed as follows:

> (1) Is the variation that provides the raw material for evolution continuous or discontinuous?
> (2) Can the mean of a continuously varying characteristic be shifted by selection?
> (3) How important is distant ancestry for the hereditary constitution of descendants?
> (4) What methods should be employed to study variation and heredity; the *actuarial* methods of biometry using randomly breeding populations constituted of individuals of unknown ancestry, but belonging to the same species, of those of Mendelism using planned crosses between individuals of known pedigree but belonging to different species or varieties, and marshaling the offspring into discontinuous categories?
> (5) What philosophical canons should govern the research; those of Machian positivism, or those more favorable to the acceptance of unobservable entities (e.g., hereditary factors) such as the Newtonian methodology of the *vera causa* principle? (Olby 1989: 300)

Weldon pioneered the actuarial approach of biometrics, following direct inspiration from Galton (Olby 1989; Pence 2011; Provine 2001; Reznick 2010). He saw the study of evolution as seeking to correlate trait variation within organisms as well as mortality rates with unusual variation in particular traits, and trait variation between parents and offspring. He also saw the need to track proportion of variants within populations. Tight correlation of trait variation could enable classification of species, and mortality rates and reproductive success associated with variants could track possible selection, in the context of population level data. As Reznick (2010) points out, Weldon repackaged Darwin's thesis but in arithmetic terms that required empirical data, painstakingly won from much measurement in the laboratory and field, and then statistics. It was this last requirement that saw Weldon persuade Pearson to collaborate and develop correlational methods.

Pearson's first foray enabled Weldon's discovery that the front breadths of a species of shore crab (*Carcinus maenas*) were not normally distributed, but were in fact bimodal within the population measured, whereas all other traits were on a normal curve (Pence 2011; Weldon 1893). Pearson mathematically supported Weldon's hypothesis, by developing curve-fitting methods, that indicated segregation into "races" of shore crab (Provine 2001). The argument made was that selection had operated on a continuously varying trait to split it, such that both peaks of the bimodal subpopulations were moving away from one another.[9]

While Galton approved of this general data led and statistical approach, both Weldon and Pearson were convinced that evolution was not reliant upon discontinuous variation as Galton asserted, and this led to friction. According to Provine's analysis, Weldon and Pearson saw Galton's interpretation of his own data about regression as in error. Both felt that Galton had failed to account for the possibility that, in a reproducing population, individuals that scored above the median value for a given trait may well breed with one another, reducing regression to the overall mean of the population to zero within this subpopulation. This argument, based on a view of reproductive ecology, was sufficient for Weldon and Pearson to dismiss Galton's argument, and to continue pursuing a continuous variation approach to evolution.[10]

While Weldon, Pearson, and Galton remained on good terms, the relationship between Weldon and Bateson that had previously been amicable and collegial broke down as their debate soured and became interpreted on both sides as personal attack (Olby 1989).[11] Galton continued to pursue his ideas and developed a law of ancestral heredity, eventually modified by Pearson, that was purely descriptive of phenotypic inheritance and thus rendered neutral with regard to predictive utility (Mayr 1982). Galton developed particulate views on inheritance but treated inheritance as blending (something which is implied in Darwin's *pangenesis* theory) such that the traits from each parent are averaged in the offspring, a mechanism that supported his view of regression. As a result, particular patterns of inheritance across generations were inexplicable under his formulation and his theory failed; and other ideas about inheritance were soon to emerge.

[9] According to Pence (2011), while Pearson assisted Weldon in developing statistical approaches to this problem, he later intervened in a critical debate about the methods used in another shore crab paper that attempted to tie traits to particular death rates (Weldon 1895). Pearson chose not to defend Weldon, but to point to the idealized nature of the approach and some fundamental problems of measurement that made the data questionable; he felt causation due to the claimed traits was not formally supportable.

[10] Weldon also felt that there was no reason given to exempt sports from regression to the median (Provine 2001).

[11] Bateson's views are discussed in the next section.

2.5.1 Mendel and Beyond

The concerns of Weldon and Galton were supplanted by the rediscovery of Mendel's classic 1866 paper (Abbott and Fairbanks 2016) by de Vries, Correns, and Tschermak in independent publications in 1900. In these papers each author laid claim to laws of inheritance that were later found to mirror those of Mendel. It is not clear that these separate publications led to precise rediscovery, and indeed de Vries and Correns may well have read Mendel prior to their own publication thus influencing their interpretations, while Tschermak if he did may not have understood it fully (Fairbanks and Abbott 2016; Mayr 1982). Nonetheless, all three came to acknowledge Mendel.

Mendel's paper was the outcome of a series of crossing experiments with peas (*Pisum*). Peas were chosen as they enabled hybrids (offspring) not of intermediate form, but that expressed a character from one or the other parent. This fitted his objective which he stated as follows (from the Darwinian translation of (Abbott and Fairbanks 2016):

> If two plants that are constantly different in one or more characters are united through fertilization, the characters in common are transmitted unchanged to the hybrids and their progeny, as numerous experiments have shown; each pair of differing characters, however, unites in the hybrid to form a new character that generally is subject to variation in the progeny. To observe these variations for each pair of differing characters and to ascertain a law according to which they occur in succeeding generations was the objective of the experiment. This experiment, therefore, breaks up into just as many individual experiments as there are constantly differing characters in the experimental plants (Abbott and Fairbanks 2016: 408).

Mendel chose seven characteristics that differed distinctly, and he rejected traits that varied "more or less" and did not permit sharp distinctions to be drawn between types. He was also aware that some characters dominate in their expression in offspring generations. Other characters were argued to be "latent" in the subsequent generation only to reappear later in further generations. Such characters were termed recessive.[12]

Mendel used 22 varieties of *Pisum* that bred true when self-fertilized. This enabled him to take individuals from these varieties that differed on one of the seven characteristics, and to cross them to produce what he termed hybrids. Those traits were then tracked across the subsequent generations. An example of one of the seven is whether or not the ripe seeds of the plant were round and smooth or deeply wrinkled. Thus, this characteristic, as with the remaining six, represented a pair of possible morphs. When self-fertilized a plant with round seeds produced offspring with round seeds, and similarly for wrinkle seeded plants. But what would happen

[12] Mayr (1982) notes that de Vries, from his own crossing work, determined characters to be either active or latent, corresponding to dominant or recessive traits in Mendel. This use of latent possibly suggests some awareness of Mendel prior to publication.

Table 2.1 Two of Mendel's crosses. The top panel represents the cross between a true breeding round seed planted (RR) and a true breeding wrinkle seeded plant (ww). Each parent contributes one particle (R or w) that conveys a trait to the next generation. In the F1 (filial 1, or first generation) all offspring have both an R and a w contribution, but all present as round seeded. The bottom panel represents the self-fertilization of an individual (Rw) from the F1 to produce F2. Here the 3:1 ratio of round to wrinkled is produced, again due to the contribution of single particles to each fertilization event

Parental 1	w	w
R	Rw	Rw
R	Rw	Rw
Parental 2	R	w
R	RR	Rw
w	Rw	ww

when a round seed plant was crossed with a wrinkly seed plant, and what would the hybrid offspring themselves produce when they were self-fertilized?

The answer to this question was that the first generation (F1) from the round and wrinkled cross was all round seeded. However, upon self-fertilization and the production of the next generation (F2) a ratio of 2.96:1 (round:wrinkled) was produced (Mayr 1982). Across the other traits similar patterns emerged creating an approximate 3:1 ratio each time.

Mendel produced an F3 generation by self-fertilizing F2 plants. The wrinkle seeded plants all bred true, but the round seeded produced approximately one third that were rounded seeded, and the remaining plants produced round to wrinkled in the 3:1 ratio.

Following these single character experiments Mendel then moved on to investigate patterns of inheritance crossing plants with reference to more than one character. In these experiments Mendel found that each character (with its two morphs) was inherited as described above, with no apparent interaction between them. This became Mendel's law of independent assortment.

The transitions between F1, F2, and F3 can be presented in matrices. In Table 2.1 two transitions are represented.

As Table 2.1 makes clear, Mendel's experiments were interpreted in particulate terms. Mendel was working well before the notion of genetic inheritance had been developed, and so cannot be credited with the development or refinement of genetic theory. But he was able to show distinct patterns of the inheritance of characters that suggested each individual contains two contributions to the kinds of traits that Mendel had specifically focused upon. During the formation of the germ cells, a parent gives just one of those contributions, or particles and the gametes later fuse with those from the other parent to form a full individual with two particles for that character. This became known as Mendel's law of segregation (Fairbanks and Rytting 2001). Moreover, those particles can be dominant or recessive to one another, such that when an R particle is present the plant will always produce round seeds. But for wrinkle seeded forms to emerge, the plant must have two recessive w particles. Such an individual is homozygous recessive, whereas round seeded plants can be either homozygous (RR) or heterozygous (Rw).

Mendel's focus was upon what can be termed discontinuous traits. There is no blending between roundness and wrinkliness in the seeds of the pea plants. It was this discontinuity, and the distinctive patterns of inheritance attributed to those traits that led to Mendel being adopted by those opposed to gradual Darwinian evolution. Mendel's work, and that of his rediscoverers, appeared to offer hope to those who took positions similar to Galton's. This included Bateson, who read Mendel's paper in May 1900 on his way to a meeting of the Royal Horticultural Society. The paper supported his general position and he immediately incorporated it to his comments at the meeting. From that point Bateson was a dedicated Mendelian, using particulate arguments about discontinuous traits to bolster views against gradual Darwinian evolution (Provine 2001).

Provine notes that what Bateson saw was a complementarity between discontinuous traits and discontinuous evolution. But he failed to notice other comments by Mendel when studying varieties of *Phaseolus*, with white and purple flowers. The hybrids from white and purple crosses were all purple, but the purple hybrids were able to produce offspring with purple, red, violet, and white forms. Mendel's interpretation of this was that overall color was under the control of perhaps two, or more color traits, and that the particles for each of those traits independently assorted and segregated just as those in the *Pisum*. Here Mendel is conceding the possibility of continuous variation, underpinned by particulate inheritance. This would make Mendel compatible with Darwin, and there is reason to think that while Darwin did not read Mendel, Mendel was influenced by Darwinian ideas (Fairbanks and Abbott 2016).

Bateson's determined use of Mendel to argue against Darwinian evolution prompted the biometricians, such as Weldon and Pearson, to take issue against Mendelian views. They are thus guilty of not properly reading Mendel, but rather of reacting to a secondary interpretation, and only dealing with those aspects of Mendel. This led to a number of controversial interactions and meetings that Provine (2001) details well.

An important engagement with Mendel was that of de Vries and his development of mutation theory, which he used to extend Darwin's ideas on pangenesis. De Vries disagreed with the Lamarckian tone of Darwin's induced variation (Sect. 2.2.2) and instead suggested that during division a *pangen* might produce an altered *pangen*, and this could be active. While different *pangens* could combine in multiple ways in the next generation this only explained individual variation, and de Vries regarded this as unimportant for evolution, siding with Galton's position on regression. However, the alteration of *pangens* during division explained the emergence of new characters, not just variation about old, and thus a form of new variation that could be passed on that sat well with discussions of *sports*. Thus, de Vries came to see evolution as discontinuous also, and found himself against the biometricians which led to a long association with Bateson. Nonetheless, his mutation idea was something of interest in its own right and laid the path for future thinking outside of the debate between Mendelians and the biometricians.

Between 1900 and 1910 Mendelians, buoyed by the experimental work of de Vries and others, argued against Darwinian selection. Thomas Huxley and Bateson felt that continuous variation was too trivial to generate significant selection pressure, while Galton continued to rely upon his regression argument. This argument was explored by Johannsen (a Danish botanist) who in 1901 embarked upon a series of more thorough experiments than Galton that took account of the detail of pure breeding lines. The outcome of this surge of experimental work became quite technical as the following extract from Provine makes clear:

> Galton himself thought that selection was ineffective in the face of regression. The deviation of the offspring of selected parents from the mean of the population was only two-thirds the deviation of the parents, and soon a balance between regression and selection must be reached. De Vries used the same argument. Pearson argued that Galton had misunderstood his own law of regression. Only if the mean of the selected parents reproduced the mean of the population would Galton's argument hold. Regression was really to the parental mean not to the mean of the population. Therefore regression did not counteract the effects of continued selection and evolution could proceed continuously. Johannsen argued, like Pearson, that Galton had misunderstood his law of regression. In a pure line, regression was complete to the type of the line and selection was ineffective against such regression. Galton observed incomplete regressions because he failed to analyze biologically his populations (Provine 2001: 95–96).

Johannsen believe that there were five types of variation (Roll-Hansen 2014):

1. Sub-species or varieties.
2. Hybrid variation.
3. Individual variation.
4. Environmentally induced variation.
5. Mutations.

It is against this list that he developed his key experiments on beans (*Phaseolus*). The beans he chose relied heavily upon self-fertilization, which meant that any hybridization through crosses was a weak source of variation within the plant. He then went on to produce three generations of bean, via self-fertilization, measuring the bean seeds produced in each generation with reference to those of their parents. In this way a number of pure lines were produced, descended each from one ancestor and he was able to analyze the third type of individual or fluctuating variation.

Johannsen's key finding was that there was no inheritance of individual variation between generations, and thus there was no correlation between the parent and the offspring (indeed he claimed, without reporting, that the correlation was zero (Provine 2001)); instead it is the *type* of the pure line that was the cause of the average character seen in offspring, something Mayr characterized as an essentialist error (Mayr 1998). From this Johannsen concluded that individual variation was not available for natural selection in pure lines, leaving only hybridization and mutation as sources. Roll-Hansen expressed Johannsen's conclusions slightly differently, suggesting that when hybridization (or recombination) and mutation are excluded then heredity is stable and there was no continuous change (Roll-Hansen 1989).

This result led to a more stringent form of Galton's view such that in a population with extreme forms of bean represented (say very large or very small seeds), those extreme forms would include some pure lines and their offspring would be determined by the *type* of the pure line, rather than by parental influence. Galton straddled the biometric and Mendelian divide. As we have seen, he believed in individual variation and spent much time measuring it in biometric terms but concluded that regression to the mean would remove the extremities required for Darwinian selection. As such he sided with *sports*, which were promptly aligned with Mendel and mutation theory. Johannsen agreed that *sports* were the better model for evolution, but he specifically argued that regression to the mean did not happen as Galton envisaged but rather only within the total population as selection between pure lines. Within a subpopulation, as modelled by Galton, Johannsen claimed it did not occur. Galton's own findings were the result of shifts in the frequency of pure line types, due to the breeding set-up, and a consequence of not clearly pursuing pure lines and hybrids. As Roll-Hansen (1989) notes, Johannsen saw this as both the championing and undoing of Galton's view.

Johannsen received swift criticism from Weldon and Pearson in 1903[13] who focused in particular upon reanalyzing the data presented and demonstrating correlations greater than zero between parent and offspring. Provine (2001) and Roll-Hansen have differing views on how effective those criticisms were, with Provine seeing them as successful defenses of the biometric position and of continuous variation positions. Nonetheless, Johannsen was taken up by the community and Mayr (1980) reports notable figures such as Morgan commenting in the 1930s that the Darwinian view of selection of the extremities of continuous variation was known to be wrong. Mayr attributes some of this to the lack of direct evidence for Darwinian views and also to the fact that many Darwinians still incorporated Lamarckian soft inheritance, and in particular notions of use and disuse, into their thinking about sources of variation.

2.6 Conclusion

Stopping at the point in history that we have in Sect. 2.5.1, in the midst of arguments about the nature of variation and selection, is to stop just prior to the beginning of the development of the Modern Synthesis. The Modern Synthesis is minimally about resolving this debate, synthesizing Darwinian and Mendelian views, and we shall turn to that in Chap. 3.

Darwin was committed to a certain kind of variation but bracketed its causes during his theory building. His view on gradual evolution had its foundation in his observations of the natural world but possibly also in metaphysical reflections (Mayr 1982) and culturally embedded beliefs (Gould and Eldredge 1977). These

[13] Johannsen published his experiments in 1903, and Weldon and Pearson published reviews in *Nature* and *Biometrika* in the same year.

commitments thus pre-dated Darwin's development of the theory of evolution by natural selection (Gould 2002) and provided a context for it.

Jenkin made argument about the parameters for continuous variation and natural selection, from a typological position about species, and this essentialism is at odds with Darwinian thinking. The thrust of Jenkin's argument was that natural selection could not create new species with continuous variation over long historical time, as Darwin had claimed. This argument relied in part on regression and thus prefigured that of Galton, and the broader biometrician-Mendelian tension. As Roll-Hansen (1989) noted, that tension was really focused upon whether selection created new forms, as Darwin claimed, or merely selected between them (see Sect. 2.4 for related discussion).

Scientists on either side of the debate invested heavily in experimental work to decide which view was correct, and they agreed on the terms and structure of those experiments even when they did not agree about interpretation of particular findings. There is some disagreement between historians as to the effectiveness of the biometricians' arguments but until the late 1930s the Mendelian view was regarded as the better one. Johannsen's final position on the role of natural selection was that:

> Selection cannot produce a deviation from the mean in self-fertilizing species, "and even the most careful experiment with cross-fertilizing plants and animals confirm most convincingly our interpretation of an inability of selection to achieve more than a mere isolation or separation of previously existing constitutionally different organisms: selection of different individuals creates nothing new; a shift of the biological type in the direction of selection has never been substantiated!" He finally concludes that it is "completely evident that genetics has deprived the Darwinian theory of selection entirely of its foundation and . . . the problem of evolution is still an entirely open question" ((Mayr 1982: 585) citing Johannsen's conclusions from 1915).

But note the phrasing, this is a criticism of Darwin's view of selection, but it is not an argument against evolution *per se*. That idea was firmly established in the scientific imagination, and as the quotation suggests, was an active research area.

References

Abbott S, Fairbanks DJ (2016) Experiments on plant hybrids by Gregor Mendel. Genetics 204 (2):407–422. https://doi.org/10.1534/genetics.116.195198

Allen GE (1969) Hugo de Vries and the reception of the "mutation theory". J Hist Biol 2(1):55–87. https://doi.org/10.1007/BF00137268

Brinkman PD (2010) Charles Darwin's Beagle voyage, fossil vertebrate succession, and "the gradual birth & death of species." J Hist Biol (Vol. 43, Issue 2). https://doi.org/10.1007/s10739-009-9189-9

Büttemeyer W (2005) Popper on definitions. J Gen Philos Sci 36(1):15–28. https://doi.org/10.1007/s10838-005-6037-2

Delisle RG (2019) Charles Darwin' s incomplete revolution: the origin of species and the static worldview. Springer Nature Switzerland

Desmond A, Moore J (1991) Darwin. Michael Joseph

Fairbanks DJ, Abbott S (2016) Darwin's influence on Mendel: evidence from a new translation of Mendel's paper. Genetics 204(2):401–405. https://doi.org/10.1534/genetics.116.194613
Fairbanks DJ, Rytting B (2001) Mendelian controversies: a botanical and historical review. Am J Bot 88(5):737–752. https://doi.org/10.2307/2657027
Gould SJ (2002) The structure of evolutionary theory. The Belknap Press of Harvard University Press
Gould SJ, Eldredge N (1977) Punctuated equilibria: the tempo and mode of evolution reconsidered. Paleobiology 2(2):115–151. https://doi.org/10.1017/S0094837300005224
Huxley JS (2010) Evolution: the modern synthesis: the definitive edition. MIT Press
Joshi A (2017) Weldon's search for a direct proof of natural selection and the tortuous path to the Neo-Darwinian Synthesis. Resonance 22(6):525–548. https://doi.org/10.1007/s12045-017-0497-7
Mayr E (1982) The growth of biological thought: diversity, evolution and inheritance. Belknap Press
Mayr E (1991) One long argument: Charles Darwin and the genesis of modern evolutionary thought. Harvard University Press
Mayr E (1998) Prologue: some thoughts on the history of the evolutionary sciences. In The evolutionary synthesis: perspectives on the unification of biology (pp. 1–48). Harvard University Press
Olby R (1989) The dimensions of scientific controversy: the biometric-Mendelian debate. Br J Hist Sci 22(3):299–320
Pence CH (2011) "Describing our whole experience": the statistical philosophies of W. F. R. Weldon and Karl Pearson. Stud Hist Philos Sci C Stud Hist Philos Biol Biomed Sci 42(4):475–485. https://doi.org/10.1016/j.shpsc.2011.07.011
Popper KR (1945) The open society and its enemies. Routledge
Provine WB (2001) The origins of theoretical population genetics, 2nd edn. University of Chicago Press
Reznick DN (2010) The origin then and now: an interpretive guide to the origin of species. Princeton University Press
Roll-Hansen N (1989) The crucial experiment of Wilhelm Johannsen. Biology & Philosophy 4 (3):303–329. https://doi.org/10.1007/BF02426630
Roll-Hansen N (2014) Commentary: Wilhelm Johannsen and the problem of heredity at the turn of the 19th century. Int J Epidemiol 43(4):1007–1013. https://doi.org/10.1093/ijc/dyu066
Ruse M (1975) Charles Darwin and artificial selection. J Hist Ideas 36(2):339–350
Schwartz JS (1974) Charles Darwin' s Debt to Malthus and Edward Blyth. J Hist Biol 7(2):301–318
Sulloway FJ (1982a) Darwin's conversion: the Beagle voyage and its aftermath. J Hist Biol 15 (3):325–396. https://doi.org/10.1007/BF00133143
Sulloway FJ (1982b) Darwin and his finches: the evolution of a legend. J Hist Biol 15(1):1–53. https://doi.org/10.1007/BF00132004
Sulloway FJ (1984) Darwin and the Galapagos. Biol J Linn Soc 21(1–2):29–59. https://doi.org/10.1111/j.1095-8312.1984.tb02052.x
van Inwagen P, Sullivan M (2020) Metaphysics. In The Stanford encyclopedia of philosophy (Spting 202). https://plato.stanford.edu/archives/spr2020/entries/metaphysics/
Van Wyhe J (2013) "My appointment received the sanction of the Admiralty": why Charles Darwin really was the naturalist on HMS Beagle. Stud Hist Philos Sci C Stud Hist Philos Biol Biomed Sci 44(3):316–326. https://doi.org/10.1016/j.shpsc.2013.03.022
Weldon WFR (1893) On certain correlated variations in Carcinus maenas. Proc R Soc B Biol Sci 54:318–329
Weldon WFR (1895) I. Report of the committee, consisting of Mr. Galton (Chairman), Mr. F. Darwin, Professor Macalister, Professor Meldola, Professor Poulton, and Professor Weldon, "for conducting statistical inquiries into the measurable characteristics of plants and anima. Proc R Soc Lond 57;360–379. https://doi.org/10.1098/rspl.1894.0165

The Modern Synthesis

3

Abstract

The previous chapter gave a history of evolutionary theory from Lamarck up until the debate between the biometricians and Mendelians. What happened from that point on marks the emergence of neo-Darwinism and then the Modern Synthesis. In this chapter I outline these transitions and interpret the Modern Synthesis as a two-phase transition in keeping with analysis from Gould. However, I claim that the period of synthesis really extended to the mid-1970s when Dawkins gave his definitive statement of gene-level selection and adaptationism. This innovation was a consequence of inclusive fitness theory. The chapter also covers multi-level selection theory, arguing that this is very much a product of the Modern Synthesis due to a reliance upon adaptation and natural selection. The chapter concludes by pointing out the use of information terms by Dawkins, a theme that will be picked up in later chapters.

Keywords

Saltation · Lamarckism · Orthogenesis · Neo-Darwinism · Modern Synthesis · Restriction · Hardening · Adaptation · Natural selection · Gene-level selection · Individual-level selection · Inclusive fitness theory · Multi-level selection

3.1 Introduction

In this chapter I discuss post-Darwinian transitions in evolutionary theory. What I present is a traditional synopsis of these transitions which maps the emergence of neo-Darwinism and then the Modern Synthesis itself. In keeping with Gould's (2002) analysis of the Modern Synthesis I present this as a two-phase transition.

I term this a traditional account because it follows the pattern that is often reflected in criticisms of the Modern Synthesis. To that end, I do this not to diminish the buoyant and very new scholarship in evolutionary studies (see, for example,

T. E. Dickins, *The Modern Synthesis*, Evolutionary Biology – New Perspectives on Its Development 4, https://doi.org/10.1007/978-3-030-86422-4_3

these two edited collections (Delisle 2017, 2021)) but rather to make clear the starting point for the debate generated by advocates for an Extended Evolutionary Synthesis. The works I am citing, and the story I am telling is a version of the one upon which these critics cut their teeth.

Within the traditional account there are multiple ideas and differences of view battling with each other across historical time, as biologists and other theorists have tried to make sense of Darwinian evolution in light of evidence and philosophical frameworks. I make comment on the nature of the Modern Synthesis, as a framework, in Chap. 9. For now, my intention is to demonstrate the complexity of that framework as an emerging theoretical object, and to show the roots of many foundational assumptions made by contemporary critics of the Modern Synthesis. Given the complexity of the Modern Synthesis, as traditionally portrayed, the counter claims from Extended Synthesis scholars are equally diverse.

3.2 Darwin Eclipsed

What Darwin revealed was a kind of change, evolution, which accounted for the diversity of form found in the natural world. His theory of natural selection provided a general abstraction:

> Darwin's scheme embodies three principles:
>
> Different individuals within a population have different morphologies, physiologies and behaviours (phenotypic variation).
>
> Different phenotypes have different rates of survival and reproduction in different environments (differential fitness).
>
> There is a correlation between parents and offspring in the contribution of each to future generations (fitness is heritable).
>
> These three principles embody the principle of evolution by natural selection. While they hold, a population will undergo evolutionary change (Lewontin 1970: 1).

This form of abstraction has been termed a general theory of evolution (Webb 2011) for it captures the properties and dynamics of a number of systems, not all of which are classified as life. In other words, this definition is neutral with regard to the substrates that deliver variation and inheritance (Haig 2007). To that end it is Darwin's positioning of evolution at the population level, and his push away from transformation toward a variational mode of thinking, that are his most distinct contributions.

In Chap. 2 we saw how the topics of inheritance and variation generated much debate, most especially between the biometricians and Mendelians. That debate led to general doubt within the biological community about the efficacy of natural selection as a source of creativity in evolution until the late 1930s. While the Mendelians argued against gradualism and the selection of continuous variation, they effectively promoted a view of saltation, of leaps in the emergence of form. What this left was a commitment to descent with modification, an older idea developed by the transmutationists which Darwin had made his own through meticulous natural history and cataloging (Pigliucci 2007).

Saltation was one of three ideas that acted to *eclipse* Darwinian theory in the late nineteenth century and into the early twentieth, to borrow Julian Huxley's term (Bowler 2017; Huxley 2010; Mayr 1982). The other two were Lamarckism, in various forms broadly packaged under the term neo-Lamarckian, and orthogenesis.

Darwinian evolution is typically seen as profoundly non-Lamarckian. Lamarck saw change as internally motivated by the organism, whereas Darwin had it as the consequence of biotic interactions (Depew 2010). However, Darwin did allow some room for the inheritance of acquired characteristics in the first edition of the *Origin*, and its role increased in subsequent editions. This has been interpreted as an attempt to find solutions to key problems brought up by the discussions around variation and inheritance by Jenkin (Koonin and Wolf 2009). Specifically, many have assumed Jenkin to have claimed that blending inheritance would make individual continuous variation unavailable to natural selection. However, Jenkin was really only applying this argument to *sports*, or saltations, and removing them from consideration (Vorzimmer 1963). According to Mayr, who followed Vorzimmer's line of argument, Darwin nonetheless became increasingly convinced that the environment had a direct effect leading to variation in organisms, and quotes him in a letter to Galton in 1875 confessing to an increased reliance upon modification by use and disuse (Mayr 1982). As discussed in Sect. 2.2.2, the role of environmental modification became enshrined in Darwin's 1868 theory of pangenesis.

Darwin regarded variation as undirected, even if external modification can lead to variation (Bowler 2017). Selection provided direction for evolution. But the Lamarckian and orthogenic approaches to evolution both permitted directed variation. In Sect. 2.1 we discussed how Lamarck introduced adaptation to evolution through the active accommodation by the organism to the environment. Those developmentally induced variants could then be inherited. One of Darwin's innovations was to make adaptation a result of natural selection operating over variation. Bowler (2017) makes clear that Lamarck's view of adaptation was a strong form of development induction—it was not the simple revelation of a hitherto unused capacity but rather the addition of an extra element, or stage to the life cycle of the organism as a result of its own actions, allowing phylogeny to be seen as recapitulated in ontogeny (Gould 1977).

Lamarck's ideas were actively adopted, in several quite different forms, by scholars who have been latterly grouped as the neo-Lamarckians. This term gives the impression of a coherent position, which is not borne out by historical analysis (Bowler 2017; Mayr 1982). Some preferred Lamarck to Darwin as his view permitted agency, a role for the organism in evolution and was therefore more compatible with certain moral views. This position later became attached to the idea of a non-material life force, a fundamental drive within organisms, enabling them to creatively engage with their own evolution. In this way some neo-Lamarckian views were aligned with *vitalist* metaphysics. This vitalism was later married to teleological notions of a divine plan that was unfolding during the course of evolution (Loison and Herring 2017).

Another use of Lamarck was in the service of social progress. Bowler (2017) has this as a reaction to the social Darwinism associated with Spencer (Becquemont 2011). Rather than allow selection to play out, a Lamarckian view would seek to engineer social conditions to enable people to acquire desirable characteristics and improve their situation.[1]

Orthogenesis, as with Lamarckian views, permits directed variation but it commits to an unfolding of a predetermined trend driven by the habits of organisms that may see variations being produced that go beyond the requirements of the environment and to even become deleterious (Bowler 2017). Orthogenesis has been seen as a teleological position on evolution (Mayr 1982) such that the variants produced along a line of transmutation were directed to some end and not subject to selection (Allen 1969). But as Bowler points out several orthogenesis theorists deliberately tried to remove teleology from their accounts, and to rely on concepts such as the laws of development while also pointing to the possible negative effects which would be unlikely to feature in some progressive plan. These views were more about constraint and path dependency. The various mechanisms and versions of orthogenesis theory that were developed were all focused upon bringing to the fore important sources of variation in organisms, caused by mechanical and development factors, that challenged the creative role of natural selection (Ulett 2014).

3.3 Synthesis

The Modern Synthesis is a term widely used to refer to the paradigm that frames modern evolutionary biology. The term was coined by Julian Huxley (2010) as the subtitle to his original and extensive review of developments in biology since the rediscovery of Mendel.

The use of *synthesis* implies a prior period of thesis and antithesis. The initial thesis was that of Darwin and his account of evolutionary change through natural selection. As described in Chap. 2, Darwin, and thus Darwinians regarded evolution as a gradual and incremental process reliant upon continuous variation. This became the hallmark of Darwinian thinking (Kutschera and Niklas 2004). While Darwin helped to firmly establish the idea of evolution within the scientific community and broader public, it was only his notion of descent with modification that initially found favor. As we have seen (Sect. 3.2), Darwin was challenged first by versions of Lamarckian theory and the concept of orthogenesis, that emerged in response to what was perceived as an overly mechanistic view of nature with little room for previous religious and metaphysical beliefs. These positions effectively challenged the role of natural selection by proposing alternative models of transmutation or even progression.

[1] A concern with social progress was not confined to Spencerian progressives nor to neo-Lamarckians. Many neo-Darwinian and Modern Synthetic thinkers also had strong views on the possibility of harnessing evolution to change society (Esposito 2011).

The rediscovery of Mendel in 1900 (Sect. 2.5) added to discontents with Darwinian theory and thus to what Julian Huxley referred to as the *eclipse* of Darwin. This was a theory of inheritance that provided both the mechanism for passing on traits but also appeared to support the emergence of new variants through saltation. Mendelians saw particulate inheritance as incompatible with Darwinian theory for several reasons. For example, Thomas Huxley, de Vries, and Bateson (who coined the term mutation) felt that continuous variations were too small to generate significant selection pressures. Galton felt that selection on continuous variation would be limited by regression to the mean and de Vries felt that large, discontinuous mutations were the source of variation and essential to evolution. As Provine (2001) documents, this division between Darwinian and Mendelian views of evolution persisted for a quarter of a century and not for entirely scientific reasons. There was much personal animosity between several leading characters on both sides of the debate and Provine argues that this delayed the inevitable synthesis. Social processes are not uncommon forces in scientific transition, and the development of evolutionary theory was no exception (Reif et al. 2000; Smocovitis 1992).

Where Darwinism was the thesis, Lamarckism, orthogenesis, and the saltation views of the Mendelians were the antithesis. However, not all the antithetical ideas were to be incorporated into the final synthetic piece. The synthesis was only between Mendelian genetics, moved on some way from the positions taken up in debate with the biometricians, and Darwinian evolution. This synthesis is often regarded as dependent upon the development of population genetics (Smocovitis 1992) but Mayr regards the removal of Lamarckism, orthogenesis, and saltationist thinking as a more important first stage (Mayr 1993). This first stage has been referred to as a period of restriction (Gould 2002)[2] and the resulting theoretical framework at this point is termed neo-Darwinism (Futuyma 2015; Kutschera and Niklas 2004). Thus, as Futuyma makes clear, there are three broad stages in the emergence of modern evolutionary theory:

1. Darwinism
2. Neo-Darwinism
3. Modern Synthesis

We have dealt with Darwinism and the debates emerging immediately after the publication of the Origin. Now we must turn to the remaining parts of this scientific transition.

[2] Smocovitis credits Provine with using the term constriction for this same period (Smocovitis 1992).

3.3.1 Neo-Darwinism

While the term *neo-Darwinism* is often applied to all post-Darwinian thinking, most historians agree with Futuyma's three-part transition to modern evolutionary theory. More particularly they also associate this second stage with the work of Weismann to show the separation of the somatic and germ lines, which he used to refute Lamarckian ideas. But, the actual term, *neo-Darwinism*, was applied to Weismann's position in the final words of a critical, and somewhat ill-tempered review: "The school of Weismann may properly be called Neo-Darwinian: pure Darwinian it certainly is not" (Romanes 1888). Romanes was objecting to Weismann's removal of the Lamarckian concept of use and disuse, while still referring to his position as Darwinian. For Romanes, Darwin's adoption of these concepts meant that pure Darwinism must include them.

Weismann argued that each body develops from a single cell within which there is germ-plasm (Haig 2007). The *ids* constitute the germ-plasm, each containing the material determinants to grow a new body. For any change to be heritable that change must be wrought in the *ids* and not in the body, for it is the *ids* that are passed on during reproduction. This argument was based on a denial of a homuncular view of generative material—that which is passed on is not a microscopic copy of the body from which it comes, but rather a set of materials that deliver an initial, developmental cause. Weismann saw heritable changes as the result of molecular and chemical changes in the germ-plasm, a kind of change very different from those more usually discussed as acquired in Lamarckian circles. It was for this reason that he rejected the role of use and disuse in the greater Darwinian scheme. The following quotation gives a sense of the kind of argument Weismann was dealing with and how he countered it:

> Detmer quotes the so-called phenomena of correlation in plants, and he believes that these instances help us to conceive how the acquired changes of the body (*soma*) of the plant may also influence the sexual cells. If the apical shoot of a young spruce fir be cut off, one of the lateral shoots of the whorl next below the section rises and becomes an apical shoot: it not only assumes the orthotropic growth of such a shoot, but also its mode of branching. The phenomenon itself is well known, and I have often observed it myself in my garden without making any botanical experiments; for this experiment is not uncommonly made by Nature herself, when the apical shoot is destroyed by insects (for example, the gall-making *Chermes*). The change of the lateral into an apical shoot occurs here in consequence of the loss of the true apical shoot, and is therefore really dependent upon it. The only difficulty is to understand how these and many other kindred phenomena can be considered to prove the transmission of acquired characters. That correlation exists between the parts of an organism, that correlated changes are not only common but nearly always accompany some primary change, has been perfectly well known since Darwin's time, and I am not aware that it has been disputed by any one. I further believe that hardly any one would maintain that it is impossible for the reproductive organs to be influenced by correlation. But this is very far from the admission that such changes would occur in the germ-cells as would be necessary for the transmission of acquired characters. For such transmission to occur it would be necessary for the germ-plasm (the bearer of hereditary tendencies) to undergo a transformation corresponding to that produced by the external influences—such a transformation as would cause the future organism to spontaneously develop changes similar to those which its

> parent had acquired. But since the germ-plasm is not an organism in the sense of being a microscopic facsimile which only has to increase in size in order to become a mature organism, it is obvious that the developmental tendencies must exist in the specific molecular structure, and perhaps also in the chemical constitution of the germ-plasm itself. It therefore follows that the changes in the germ-plasm which would be required for the transmission of an acquired character must be of an entirely different nature from the change itself acquired by the body of the parent plant: and yet it is supposed that the former is produced by the latter as a result of correlation (Weismann 1889: 402)

Weismann also marshalled evidence of forms that simply could not have been inherited due to use and disuse, a strategy he used in particular when attacking Spencer's Lamarckian views (Weismann 1893).

Haig (2007) notes what is at stake for Weismann. Use and disuse can impact upon the adaptation of the organism to the environment—this appeared clear to him—but use and disuse cannot impact upon heredity. So, Weismann's view of adaptation was a broad one about the fit of the organism to the environment. But he saw that ability to develop a fit as a capacity inherent in the organism. Thus, there was variation in plasticity, and that variation was inherited via the germ-plasm. Those organisms that could successfully adapt themselves through plastic means to the environment were available for selection, but what was inherited was the capacity to actively adapt in that way, not the final phenotypic outcome. Weismann puts it bluntly when taking issue with the classic Lamarckian trope of the Giraffe's neck:

> The perfection of form of an organ does not however depend upon the amount of exercise undergone by it during the life of the organism, but primarily and principally upon the fact that the germ from which the individual arose was predisposed to produce a perfect organ. The increase to which any organ can attain by exercise during a single life is bounded by certain limits, which are themselves fixed by the primary tendencies of the organ in question. We cannot by excessive feeding make a giant out of the germ destined to form a dwarf; we cannot, by means of exercise, transform the muscles of an individual destined to be feeble into those of a Hercules, or the brain of a predestined fool into that of a Leibnitz or a Kant, by means of much thinking. With the same amount of exercise the organ which is destined to be strong, will attain a higher degree of functional activity than one that is destined to be weak. Hence natural selection, in destroying the least fitted individuals, destroys those which from the germ were feebly disposed. Thus the result of exercise during the individual life does not acquire so much importance, for, as compared with differences in predisposition, the amount of exercise undergone by all the individuals of a species becomes relatively uniform. The increase of an organ in the course of generations does not depend upon the summation of the exercise taken during single lives, but upon the summation of more favourable predispositions in the germs (Weismann 1889: 84).

In these writings, Weismann was committed to natural selection as the only creative force in evolution and his arguments were cogent and influential, as Mayr notes, forcing all in the field to take a position on Lamarckism (Mayr 1982). His treatment of acquired variation that relied upon germ-plasm theory not only did away with Lamarckism but also orthogenesis, as both were concerned to pursue sources of variation that Weismann saw as outside of natural selection's grasp.

3.3.2 The Modern Synthesis: Phase I

The synthesis was due in large part to the mathematical modelling of the population dynamics of gene pools by Fisher, Haldane, and Wright, which saw the emergence of population genetics. Evolutionary forces could account for changes in gene frequencies within populations such that mutations were one key source of new variation and natural selection a principal method of sorting variants. Mendelians and Darwinians were effectively discussing the same phenomena but at different levels of explanation (Dickins and Rahman 2012; Scott-Phillips et al. 2011). Gould (2002) has referred to this as the first phase of the synthesis which he labelled a phase of restriction in which the focus of evolutionary biology was tightened. The second movement, according to Gould, was a phase of hardening, in which adaptationism and natural selection dominated the explanatory landscape.

3.3.2.1 Fisher and Haldane

Perhaps Fisher's most significant contribution to evolutionary theory was his intervention between the biometricians and the Mendelians. Fisher was a committed Darwinian, and therefore gradualist, from an early age (Depew and Weber 1996; Provine 2001) but he was also convinced that Mendelian ideas were not in opposition to Darwinian gradualism. He made his case by developing a statistical methodology, very much in keeping with biometric ethos, under the influence of the new, and probabilistic scientific models developed in physics, notably by Boltzmann (Depew and Weber 1996).

Fisher specifically took task with Pearson's claim that correlations between relatives could not be possible under a Mendelian scheme (Fisher 1918). In opening remarks Fisher corrected the common usage of standard deviations for the measurement of variability and then proceeded to statistically analyze all the components of variance within a Mendelian population. Fisher noted that while correlations between relatives for a given trait could be derived, the temptation to attribute the remaining variance to externalities was possibly in error. Thus, a correlation in height between a parent and child of $r = 0.5$ should not be interpreted as meaning that only 50% of the variance is a result of ancestry. Fisher's analysis enabled a separation of effects due to environmental factors but also of effects due to what he termed additive and non-additive heritable variation (Depew and Weber 1996). These non-additive effects were dominance relations and epistasis (or interactions between genes). Fundamentally, Fisher was arguing that for any given trait very many genes were often involved, and the differences between those genes were small and enabled continuity of variance in expressed traits. This view is the statistical extension of Mendel's own view given when discussing flower color derived during crosses in *Phaseolus* (Sect. 2.5.1). While Fisher was most probably unaware of this, in spite of detailed analysis of Mendel (Fisher 1936), he made the same observation in mathematical terms and went on to derive the regression curves that Galton used in arguments against Darwinian biometricians.

Depew captures the innovation of Fisher, from his papers in 1918 and 1922 (Fisher 1922), as follows:

> By abandoning the crude assumption that there is one gene for one trait he was able to use statistical analysis not simply to measure transgenerational phenotypic change in populations, like the biometricians from whose ranks he had risen, but to picture, describe, and define with mathematical precision the dynamics of Mendelian populations. Arrays of different genotypes, Fisher argued, are very much like arrays of atoms in the statistical-mechanical theory of gases. Any system of freely interbreeding alleles, like randomly colliding atoms in contained gases, will (counterfactually) remain macroscopically in equilibrium generation after generation until a force exerted somewhere inside the system changes its state.... Natural selection at the trait-environment interface is such a force. Its effect is to change the relative frequencies of genotypes (Depew 2017: 43).

Fisher moved evolutionary thinking away from the trait level, or that of the expressed phenotype, and into the gene level with a focus upon genetic variants, or alleles.

Fisher's most comprehensive statement came in *The Genetical Theory of Natural Selection* (Fisher 1930). Here he argued that Darwin's view of blending inheritance reduced variation and thus a great deal of new variation must be added with each generation for natural selection to work. Mendelian inheritance removed this problem because it retained variation across the generations:

> In view of the close analogy between the statistical concept of variance and the physical concept of energy, we may usefully think of the heterozygote as possessing variance in a potential or latent form, so that instead of being lost when the homozygous genotypes are mated it is merely stored in a form from which it will later reappear. A population mated at random immediately establishes the condition of statistical equilibrium between the latent and the apparent form of variance. The particulate theory of inheritance resembles the kinetic theory of gases with its perfectly elastic collisions, whereas the blending theory resembles a theory of gases with inelastic collisions, and in which some outside agency is required to be continually at work to keep the particles astir (Fisher 1930: 11).

For Fisher, this property of retaining variation was equivalent to the conservation laws in physics (Depew and Weber 1996). Alongside this Fisher sought to develop a model for the rate of change in a system, which Depew and Weber claim was to be the equivalent of Newton's second law, where force is the product of mass and acceleration. Thus natural selection was to be understood in proportion to other properties. This resulted in his fundamental theorem of natural selection which relied upon the concept of *fitness* understood as comparative reproductive rate. Fisher expressed the theorem as "the rate of increase in fitness of any organism at any time is equal to its genetic variance in fitness at that time" (1930: 35).

Each gene could contribute to the gene pool in the subsequent generation and the value of that contribution was the fitness of the gene. This is directly related to the shift from evolution as a process in which whole organisms are the focus of selection to one where evolution is regarded as change in gene frequencies within gene pools (Mayr 1982).

The rate of increase in fitness equated to the rate of natural selection, and this rate is dependent upon the amount of additive genetic variation in the population of alleles. So, as the amount of variation increases so does the rate of natural selection. What natural selection delivers is increased fitness, which will keep on increasing until all available variation is used. At that point the population of alleles will attain a Hardy–Weinberg equilibrium until new variation emerges.[3]

The interpretation of Fisher's fundamental theorem presented thus far is the traditional one (Frank and Slatkin 1992). But as Frank and Slatkin argue, it is in fact the case that natural selection only drives fitness up under specific circumstances. There are occasions when it can drive fitness down, such as when selection acts on two linked loci that have epistatic effects. Frank and Slatkin list epistasis, linkage disequilibrium, and frequency dependency as some properties of a genetic system that can counter Fisher's theorem.

Frank and Slatkin examine more detailed interpretations of Fisher and claim that while he did understand the role of natural selection as above, he did not see natural selection as evolution. Evolution was the result of natural selection operating over alleles but also in the context of a deteriorating environment (to use Fisher's own phrase) and thus Fisher's equations contained terms for both changes in fitness due to selection but also due to environmental cause. For Fisher, natural selection drove fitness upward, but this effect was counteracted by extrinsic environmental cause, bringing fitness toward zero. This partitioning is clarified through a discussion of work on clutch size in birds. While there is much genetic variance associated with clutch size, potentially enabling natural selection to drive clutch size up and increase fitness, clutch size remains fairly stable across generations. One reason for this is that there is competition for available and limited resource to feed any chicks, and this counters selection on clutch size. It is this focus upon equilibria within allele-space, rather than upon evolutionary dynamics per se, that Frank and Slatkin argue Fisher was trying to capture (Svensson 2016).

Haldane made a contribution to gene-linkage early in what was a diverse career (Depew and Weber 1996; Provine 2001). Following a forced intermission by World War I, Haldane then applied himself to the mathematical analysis of population genetics. Depew and Weber align Haldane with Fisher to a large extent:

> Among Haldane's and Fisher's most important conclusions were these. It does not take much mutation for natural selection to work, so mutation is not the primary agent in evolutionary change. Indeed, very small mutation rates can have significant effects on the distribution of gene frequencies, even where selection pressures are quite weak, because useful variation exists in populations, stored, for example, on the recessive side of heterozygotes. It is, in fact, very difficult to extinguish a recessive gene from a population fully. Moreover, the random or chance extinction of new genes in a finite population is rare. Since vast quantities of variation are available, therefore, and new gene combinations are constantly arising through sexual recombination, adaptation is not a matter of occasional purification but a constant and creative process, in which variation serves as fuel for natural selection (Depew and Weber 1996: 237).

[3]This equilibrium was independently discovered by Hardy and Weinberg (Edwards 2008). Its common expression is $p^2 + 2pq + q^2$, where p is the frequency of dominant alleles and q the frequency of recessive alleles. This equilibrium represents the possible crosses, delivering homozygosity and heterozygosity. The argument is simply that a population that is undisturbed by mutation, migration, and selection will retain this equilibrium state.

Haldane's most famous support for a gene frequency view was his analysis of the peppered moth findings. Dark morphs outnumbered pale in the industrial regions of Manchester in England, and eventually replaced them. The argument was that pale morphs were more easily predated against the backdrop of tree bark with heavy soot deposits from factories.[4] Haldane was able to add a mathematical model of gene frequencies to this account, claiming that the dominant dark morph had a 50% greater fertility rate than the recessive form (Provine 2001). In this account, Haldane departed company with Fisher by focusing upon a single gene explanation for the evolution of a trait.

3.3.2.2 Wright

Fisher's work was predicated upon the assumption of panmictic (randomly interbreeding) and large populations to permit all possible combinations of alleles. One reason for this was to avoid genetic drift. Genetic drift is the biological equivalent of a sampling error in science, which allows overrepresentation of types due to drawing down too small a sample from the overall very large population. More precisely put, the overall population will have a distribution of alleles, but pulling a small subpopulation from it can readily create a population that diverges from that distribution. Thus, a small interbreeding population, separated from the larger whole, can easily cause particular genes to go to fixation in the absence of natural selection simply as a result of this sampling error.

It was Wright who first saw an important role for genetic drift in real biological populations. Indeed, Wright was interested in ecologically real populations as opposed to Fisher's idealized approach, seeing them as often small and isolated with little genetic flow between them (Depew and Weber 1996). Further to this, Wright also knew from his own experimental work that the kind of large scale and free interbreeding that Fisher relied upon could very easily lower fitness simply by allowing many less useful genetic combinations to occur.

A small population, or deme, within an overall species could represent a skewed genetic sample relative to the total distribution represented in the species. If such demes were fully isolated, then they would be inbreeding which could in turn enable deleterious alleles to accumulate across subsequent generations and lead to extinction. Nonetheless, beneficial variation could also establish itself and if some individuals migrated from their deme into other demes, then natural selection could take hold of those variants bringing them to fixity within the overall population. In this way Wright proposed two processes at work, that of inbreeding and that of migration, which like Darwin were drawn from analogy with artificial selection or breeding regimes in which a balance of inbreeding and out-crossing were engineered to produce improved breeds (Depew and Weber 1996).

Wright's position was that in large populations natural selection would dominate, but in small demes genetic drift would be more powerful. However, it was the

[4]The story has become more complex in recent years as methods have improved (Cook and Saccheri 2013).

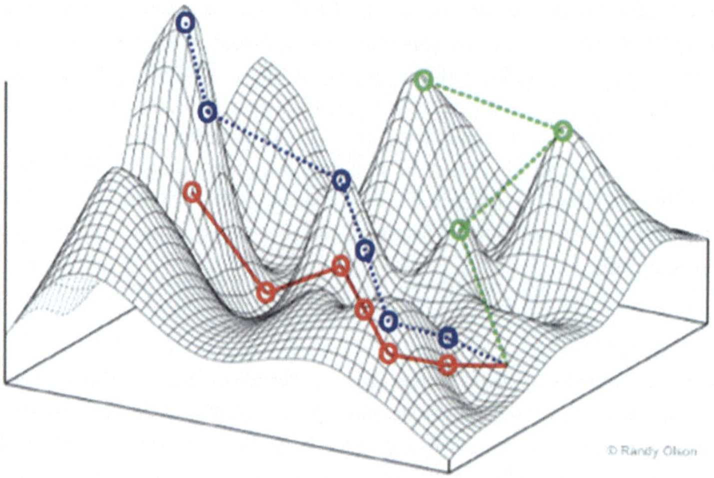

Fig. 3.1 A visualization of an adaptive or fitness landscape. The Z (vertical) axis represents fitness. The other axes might represent genes that contribute to the genotype. The colored paths on the landscape are possible paths that populations could follow during evolution. (©Randy Olson, CC BY-SA 3.0 (Figure freely available at https://commons.wikimedia.org/wiki/File:Visualization_of_two_dimensions_of_a_NK_fitness_landscape.png)

intermediate sized populations that would enable drift, migration, and selection to work together to produce evolution (Joshi 1999; Svensson 2016).

Famously Wright developed the notion of the adaptive landscape to frame and clarify his contribution (Wright 1932). An adaptive[5] landscape captures the relationships between genes (or phenotypic traits) and fitness, attempting to capture their vector like status. Fitness would increase up a z-axis, while other axes would represent different alleles within a population. Thus, where z is the dependent variable, genetic effects upon it can be registered by surveying the emergent topology or surface (Svensson 2016). This surface would be characterized by peaks and valleys, of high and low fitness, respectively, and populations would evolve by climbing to the top of a high fitness or adaptive peak (Fig. 3.1).

For Fisher natural selection could lead a population to slowly climb a peak, gradually one allele at a time (Depew and Weber 1996). Wright was aware that alleles did not operate in isolation, that they were packaged within organisms, with epistatic interactions and within population structures, and that the adaptive peak for one gene might not be the adaptive peak for another. In combination these genes might deliver different effects and different fitness values and thus climb to different

[5]The use of adaptive is meant to capture the idea of evolutionary processes adapting populations to environmental contingencies, and thus there is something plastic about this process. Coyne et al. (1997) see Wright's explanations as accounts of the emergence of adaptations which is in keeping with this view. But it should be noted that the etymology at work here does not permit the common error of talking about a trait as adaptive, when what is really meant is that it is an adaptation. An adaptive trait means a plastic trait, and might include behavioral and other physiological traits, that can adapt to circumstances albeit within a finite range of possibilities. Such a trait may or may not be an adaptation.

peaks (Dietrich and Skipper 2012). Between all these peaks were nonadaptive valleys, with fitness costs.

Wright's reliance upon natural selection, drift, and migration could explain how populations would shift between peaks and avoid the dire consequences of ending at the bottom of a valley: this was his shifting balance theory (Coyne et al. 1997). For Wright, the core problem of evolution was how populations shifted between peaks. Coyne and colleagues laid out three phases of shifting balance theory. Phase 1 relies upon genetic drift within demes initially reducing fitness and shifting those demes into valleys. But demes begin to cross to new peaks as new variation begins to accumulate due to the arrival of beneficial mutations and migrants. Phase 2 sees natural selection within demes that brings them to the top of the new adaptive peak. Phase 3 sees competition between various adaptive peaks such that some peaks win out and spread through the whole population. Wright saw these processes as delivering evolutionary stability, whereas Fisher felt that the more alleles there were in a population the greater the number of combinations there were and the fewer peaks there would be. Indeed Fisher saw such landscapes as dominated by a single ridged peak ascended by mutation and selection (Dietrich and Skipper 2012).

Provine critically evaluated Wright's adaptive landscapes, claiming that Wright made two distinct kinds of interpretation, one focused upon the genotype and the other upon the population (Skipper and Dietrich 2012). In the former, as Skipper and Dietrich relate, the axes of the landscape represent different genotypes, that can be ranked in terms of fitness along the z-axis. But these axes of gene combinations have no way of being read as there are no gradations, no scale, from which to derive the smooth surfaces Wright employed. The second interpretation has it that each point on the surface represents a mean population fitness value. In this scenario on a 0–1 gradation for gene frequency on each axis a surface can be derived but one cannot relate the fitness values directly to gene frequencies. But as Wright himself (Wright 1988) and others have objected, this is to misrepresent the heuristic role of the adaptive landscape, which was developed in 1932 in order to convey an underlying mathematical model of population genetics to a non-mathematical audience (Skipper and Dietrich 2012; Wright 1931). The heuristic value of adaptive landscapes is perhaps evidenced by the different ways in which Wright's work was rapidly adopted. Dobzhansky, Simpson, and Waddington all used adaptive landscapes to develop the detail of shifting balance theory, speciation, and population genetics, respectively (Skipper and Dietrich 2012). Skipper and Dietrich argue that the various criticisms of Wright that have emerged since (e.g., Coyne et al. 1997, 2000; Gavrilets 1997; Kauffman 1993) have all been based on differing conceptions of the underlying gene interactions leading to different kinds of topology. Thus, the difference between these critics and Wright is to do with fundamental theory, but they all end up using some version of his visual metaphor. This demonstrates the utility of the model.

Metaphors to one side, the core theoretical contribution of Wright was his shifting balance theory.

3.3.2.3 From Phase 1 to Phase 2 of the Synthesis

As noted, Gould has discussed the transitions toward the Modern Synthesis in terms of restriction and hardening (Gould 2002). The emergence of neo-Darwinism and the development of population genetics, which I have merely outlined above with a traditional focus upon Fisher, Haldane, and Wright, was the period of restriction. The focus of evolutionary theory was drawn away from saltation and orthogenesis by genetic gradualism and the power of natural selection operating over random variation. Evolutionary accounts now had to make sense in genetic terms, but as Gould notes there was a pluralism about the role of selection and genetic drift and other random phenomena. Much of the work following Fisher, Haldane, and Wright was about bringing population genetics into contact with various areas of biology, including Dobzhansky's efforts to demonstrate genetic variation in nature, Mayr's work on systematics and also on speciation, and Simpson's alignment with paleontology (Mayr 1982; Pigliucci 2007). However, in the final phase of the synthesis natural selection was, according to Gould, afforded prime place as the creative force of evolution, and this was linked to the adoption of what he terms adaptationism, where phenotypes must be "analyzed as problems in adaptation" (Gould 2002: 505).

Gould (2002) analyzes the hardening in the work of Dobzhansky, Mayr, and Simpson, showing a gradual shift to the primacy of natural selection and adaptationism from the 1940s and through into the 1960s. While some of this was a result of empirical results gathering momentum for natural selection and adaptationist hypotheses he also gives some credence to Smocovitis' view that this may have been a social response to World War II, such that evolution by natural selection was linked to progress and improvement in a way that random models of drift simply could not be (Smocovitis 1996).[6]

3.3.3 The Modern Synthesis: Phase II

Williams' 1966 book, *Adaptation and Natural Selection*, is perhaps the clearest statement of adaptationist thinking emerging from the hardening phase of the Modern Synthesis (Williams 1996). Williams opens his book with a direct reference to the debates between the biometricians and Mendelians, and he declares that the population genetics of Fisher, Haldane, and Wright decisively demonstrated that Darwinian natural selection was possible in genetic systems. Nonetheless, Williams also references more recent "overt and cryptic" (1996: 3) modern opposition to natural selection. He believes that much of this is due to a failure of imagination, noting how hard it is to imagine "how the blind play of genes could produce" humans (1996: 4). But he also sets his book against modern qualifications and additions to Darwinian natural selection "such as genetic assimilation, group

[6]More strongly Smocovitis links Modern Synthetic thinkers to Enlightenment ideals throughout her book.

selection, and cumulative progress in adaptive evolution" (1996, 4).[7] This view is written around 1963, according to Williams' note in the original preface, the same period in which Gould (2002) locates the changed, hardened views of Mayr and others. If Gould is right in his analysis, Williams might have been sensitive to the transition and seen his role as a Lockean underlabourer.[8] To address this task Williams declared his approach, telling us that his book:

> (A)dvocates a ground rule that should reduce future distractions and at the same time facilitate the recognition of really justified modifications of the theory. The ground rule—or perhaps *doctrine* would be a better term—is that adaptation is a special and onerous concept that should be used only where it is really necessary. When it must be recognized, it should be attributed to no higher a level of organization that is demanded by the evidence. In explaining adaptation, one should assume the adequacy of the simplest form of natural selection, that of alternative alleles in Mendelian populations, unless the evidence clearly shows this theory does not suffice (Williams 1996: 4–5).

He further declared a terminological convention:

> Whenever I believe that an effect is produced as the function of an adaptation perfected by natural selection to serve that function, I will use terms appropriate to human artifice and conscious design. The designation of something as the *means* or *mechanism* for a certain *goal* or *function* or *purpose* will imply that the machinery involved was fashioned by selection for the goal attributed to it. When I do not believe that such a relationship exists I will avoid such terms and use words appropriate to fortuitous relationships such as *cause* and *effect* (Williams 1996: 9).

This comment clearly flags Williams' meticulous intentions and methods in developing argument and introduces the notion of apparent design that adaptationist thinkers have since adopted (Dennett 1995). But it also makes clear his view that looser locutions have led to much confusion and error. In this spirit Williams begins to lay out criteria for considering something an adaptation. Adaptation is to be associated with "evidence of complexity and constancy" (1996: 10) and he gives the example of the lateral line of fishes that "shows a structural constancy within taxa and a high degree of histological complexity" (1996: 10). These facts were known for a long time, but what was missing was a functional account. That emerged after much morphological and physiological work, all conducted under the assumption

[7]Note the use of adaptive here. In keeping with Wright, Williams is using it to apply to evolution as a plastic, responding process over time. But later in the book Williams uses adaptive to mean adapted, which introduces a possible conflation between adaptation and plasticity. Not all adaptations are plastic.

[8]John Locke's 1689 *An Essay Concerning Human Understanding* opens with an epistle to the reader in which he states "The Commonwealth of Learning is not at this time without Master–Builders, whose mighty designs, in advancing the Sciences, will leave lasting Monuments to the Admiration of Posterity; But everyone must not hope to be a Boyle, or a Sydenham; and in an Age that produces such Masters, as the Great __ Huygenius and the incomparable Mr. Newton, with some other of that strain; 'tis Ambition enough to be employed as an Under–Labourer in clearing Ground a little..."

that the complexity and constancy indicated an adaptation—thus these criteria guided research and discovery. Here Williams has adopted an ontological commitment to natural selection and applied it as an epistemic tool—this is a clear expression of what might be termed an adaptationist program of research. But he was also clear, in keeping with the above quotation, that the concept of adaptation should not be used when less "onerous principles, such as those of physics and chemistry or that of unspecific cause and effect, are sufficient for a complete explanation" (1996: 11).

Williams clarified the concept of natural selection noting that whatever is selected must have a high degree of permanence and a low rate of change. Thus, a phenotype is too short lived to allow cumulative change across generations. This is because phenotypes are the outcome of genotype interactions with the environment and are ultimately idiosyncratic to the developmental history of the individual. Similarly, genotypes are transitory aggregates of genes, contributed by each parent in sexually reproducing organisms. In asexually reproducing, clonal species, Williams doubted that a clone would persist for long enough to have significant evolutionary effect.

> It is only the meiotically dissociated fragments of the genotype that are transmitted in sexual reproduction, and these fragments are further fragmented by meiosis in the next generation. If there is an ultimate indivisible fragment it is, by definition, "the gene" that is treated in the abstract discussions of population genetics. Various kinds of suppression of recombination may cause a major chromosomal segment or even a whole chromosome to be transmitted entire for many generations in certain lines of descent. In such cases the segment or chromosome behaves in a way that approximates the population genetics of a single gene. In this book I use the term gene to mean "that which segregates and recombines with appreciable frequency." Such genes are potentially immortal, in the sense of there being no physiological limit to their survival, because of their potentially reproducing fast enough to compensate for their destruction by external agents. They also have a high degree of qualitative stability. Estimates of mutation rates range from about 10^{-4} to 10^{-10} per generation. The rates of selection of alternative alleles can be much higher (Williams 1996: 24)

Here we have a clear statement of gene-level selection that Dawkins famously captured when summarizing the work of Williams, and others in *The Selfish Gene* (Dawkins 1976). Dawkins' simple claim was that because individual survival is central to gene survival then where we might say that "individuals act to propagate their genes" we could equally state that "genes enable individuals to propagate genes." To draw out this equivalence he developed the vehicle-replicator distinction which extended the points made by Williams above.

Genes are replicators, and the bodies they build are vehicles, but the replicator-vehicle distinction is designed as an abstraction that permits other systems to exist that cohere to the core properties associated with replicators and vehicles.[9] Replicators, then, have the property of *copying fidelity*, *fecundity*, and *longevity*.

[9]Dawkins' own example of a non-genetic replicator is that of memes, ideas, or concepts that replicate within cultural space as they are transmitted from brain to brain within the population.

They are the fundamental units of natural selection because they can survive and form lineages of identical copies. Mutation of a replicator enables new variation, but evolutionary change is only possible because of the general stability of replication: novel mutations will be faithfully copied just so long as their effect is not deleterious. DNA clearly has all these properties and Dawkins argued that genes are functionally immortal in that the data they contain is capable of transmission across multiple generations, only being ended by mutation or extinction.

Bodies are vehicles for the genes they contain, and they facilitate replication through reproduction: replication and reproduction are distinct processes, the latter enabling the former. Bodies, and the bodies they produce, do not have the properties of longevity or fidelity unlike the DNA they contain and transmit across generations. Thus, vehicles are *short lived*, relative to replicators, and their reproduction produces individuals that are not identical to the parents. But vehicles interact with the external world in a manner that facilitates genetic transmission and genes contribute to the construction of vehicular traits. Those traits that permit successful genetic replication and future representation in the gene pool are adaptations.

The title of Dawkins' book refers to the idea that genes can be modelled as self-interested, pursuing their replication through the vehicles that contain them. This should not be taken to imply that genes necessarily contribute to the construction of selfish individuals. It is possible that selfish behavior at the individual, or vehicular level is useful and will be selected, but it is also possible cooperation at that level is useful and will be selected. Nor does the selfish gene perspective imply that genes are blueprints, a set of instructions to build the vehicles. Dawkins, again taking his lead from Williams' comments on the production of the phenotype, explicitly discusses genes as catalysts that only directly influence protein synthesis and indirectly influence products further downstream from the initial protein. He argued that the "gene determines a protein sequence that influences *X* that influences *Y* that influences *Z* that eventually influences the wrinkliness of the seed or the cellular wiring up of the nervous system" (Dawkins 1989: 240). This is a long causal chain leading to an eventual phenotypic effect and one that relies upon the stable contextual factors of *X*, *Y*, and *Z*, systemic features of a developing organism that take the specific inputs from specific genes (Oyama 2000). Individual development relies, under this view, upon interrelations between different levels of biological organization.

With these principles clear, Williams then discussed how the mean phenotypic effect of a gene upon fitness is core to natural selection. Very simply, if individuals containing a particular gene, A (to use Williams' notation), replace themselves at a higher rate than those containing A', then all the individuals with A would be more fit, by definition. Thus, the maximization of individual fitness is "the most reliable phenotypic effect of selection at the genic level" (1996: 26). In passing he noted the complexities of genes selected for the reproductive benefit of close relatives, referencing Hamilton's work (Hamilton 1964) which contributed to the development of inclusive fitness theory (Frank 1998; West and Gardner 2013). Under this view we can assume that organisms are acting to maximize their average lifetime inclusive fitness, the average of the sum of direct fitness (reproduction) and indirect fitness (the reproduction of relatives). Inclusive fitness, then, captures all possible ways in which genes can be replicated across generations and enables biologists to model selection at the level of the individual. A key entailment of inclusive fitness theory is that interactions between relatives can benefit gene replication, such that a cost borne by one individual might benefit copies of the same genes within a related individual and lead to positive selection.

Thanks to Hamilton's insights, the components of social behavior can be ordered in terms of fitness benefits to the actor and to the recipient (West et al. 2007). Where both benefit, we can refer to *mutualism*. Where both are affected negatively, we can refer to *spite*. Behavior that is costly to the actor, but beneficial to the recipient is *altruistic*. *Selfish* behavior is that which is beneficial to the actor, but costly to the recipient. Where we see altruism, we might expect it to be between relatives, leading to kin selection. Kin selection captures the idea that the more closely related individuals are, the greater the indirect fitness pay off because of a large proportion of shared genes (or more technically alleles).[10] Hamilton (1964) mathematically expressed the relationship between costs, benefits, and relatedness as follows:

$$C_a < r_{ab} B_b \tag{3.1}$$

where *C* is the cost, *B* the benefit, *r* relatedness, *a* the actor, and *b* the recipient. This equation enables us to think about evolutionary dynamics and is commonly referred to as Hamilton's Rule. Altruism will only stabilize in a population whenever the cost of altruism is offset by the indirect fitness benefits of assisting a relative. Hamilton's Rule also extends our perspective beyond the direct ancestry of kin selection. Individuals can share a common gene, because of population structure and reproductive ecology, without having to be directly related, as Williams made clear when discussing the relationship of genes to phenotypes. All that is required is statistical association between copies of genes within a population and an impact upon fitness. The number of shared genes will increase the more direct ancestry is, and this property of a population can be referred to as genetic viscosity where populations of direct relatives are highly viscous.

Hamilton's is not a complete account because altruism can arise between non-relatives. Using game theory, Trivers demonstrated how altruistic behavior at one point in time could be understood as one transition in a mutualism extended over time (Trivers 1971). That extended mutualism was structured as a reciprocal tit-for-tat interaction that enabled a form of punishment to establish itself such that reciprocal exchanges of benefits would maximize individual inclusive fitness and the withdrawal of reciprocity would be costly (Axelrod 1990; Axelrod and Hamilton

---

[10] Haldane is thought to have quipped that he would jump into a river to save two brothers or eight cousins, because brothers have a 50% genetic investment in one another (coefficient of relatedness = 0.5), whereas cousins have only a 12.5% stake (coefficient of relatedness = 0.125). It was Wright who developed the mathematics of the coefficient of relatedness, but he never applied it to this problem (Harman 2010).

1981). These models are among many now in existence, all based on the same fundamental assumption of genetic natural selection as outlined by Williams.

One of Williams' principal targets in his book was that of group selection, and specifically the form of group selection championed by Wynne-Edwards in order to explain cooperative behavior (Wynne-Edwards 1962). Williams marshalled inclusive fitness perspectives against this thesis, and the debate continued beyond its publication (Sober and Wilson 2011; West et al. 2007; Wilson 2008).

Wynne-Edwards focused upon resource exploitation, imagining two groups: one in which individuals selfishly over-exploited resources and the other in which they regulated their use of resource. Exploitation was achieved through birth rate, such that the first group reproduced at the maximum rate and the second below this rate. The reduced birth rate of the second group saw it favored by selection, while the first group went to extinction. This model is now referred to as a classic group selection model because it focuses upon the survival and extinction of different groups and has been contrasted with gene- or individual-level selection. This is not to say that classic group selectionists deny gene- or individual-level selection, but they entertain the notion of selection happening at multiple levels. This idea has been more broadly captured as multi-level selection (MLS). One possibility of MLS is that selection at each level can differ in direction such that a group may be positively selected for while individuals within that group are negatively selected against.

Multi-level selection theory recognizes two versions: MLS1 and MLS2 (Damuth and Heisler 1988).

> In multi-level selection [1]:
>
> 1. "Group selection" refers to the effects of group membership on individual fitness.
> 2. Fitnesses are properties of individuals.
> 3. Characters are values attributed to individuals (including both individual and contextual characters—see below).
> 4. Populations consist of individuals, organized into groups.
> 5. Explicit inferences can be made only about the changing proportions of different kinds of individuals in the whole population (the meta-population).
>
> In multi-level selection [2]:
>
> 1. "Group selection" refers to change in the frequencies of different kinds of groups.
> 2. Fitnesses are properties of groups.
> 3. Characters are values attributed to groups (including both aggregate and global characters).
> 4. Populations consist of groups, composed of individuals.
> 5. Explicit inferences can be made only about the changing proportions of different kinds of groups in the population. (Damuth and Heisler 1988: 410)

In MLS1, differences between groups can lead to evolutionary change, but the fitness of a group in these scenarios is regarded as the sum of the fitness of each of the individual members of that group. MLS2 measures group fitness in terms of the number of offspring groups. Wynne-Edwards model was focused upon the differential survival of groups and as such was an MLS2 approach. Reproduction was of groups, and those groups containing more co-operators (understood in terms of their reduced exploitation of resources via birth rate reductions) would outcompete those with fewer or none. Wynne-Edwards saw MLS2 as a major force of natural selection, explaining the emergence of traits such as cooperation.

MLS2 or classic group selection suffers from two problems. First, animal groupings are transient, leading to gene flow between groups as and when they form (Williams 1996). Very low rates of migration, far lower than naturally seen, have been shown to destabilize classic group selection (Maynard Smith 1976). Second, empirical evidence suggests that groups reproduce at the maximum rate without reduction, thereby questioning the proposed mechanism for cooperatively managing resources. These criticisms of classic group selection proved reasonably fatal, but as seen MLS1 takes a different approach and is sometimes referred to as new group selection (West et al. 2007; Wilson 1975).

Wilson's (1975) MLS1 model contains two types of individual, altruist and selfish:

> They assort in groups for part of their life cycle, during which fitness-affecting interactions take place, before blending into the global population and reproducing. Within each group, altruists have lower fitness than selfish types. But groups containing a high proportion of altruists have a higher group fitness, that is contribute more individual offspring to the global population, than groups containing a lower proportion. So within-group selection favours selfishness, while between-group selection favours altruism; the overall outcome depends on the balance between the two selective forces. Wilson's model is thus designed to explain the changing frequency of an individual trait—altruism—in the overall population (Okasha 2006: 56–57).

West and colleagues criticized Wilson's model arguing that it was formally equivalent to inclusive fitness theory and thereby did not represent a novel theoretical approach (West et al. 2007). In the model altruism is favored in the second generation between groups because groups with more altruists have higher fitness. As a result, variance between groups is higher than that within groups. This is mathematically equivalent to assuming that relatedness has increased within groups and this is therefore an expression of standard inclusive fitness theory.

MLS1 has received increased attention in the twenty-first century, in part due to application of what is referred to as the Price Equation. This is a mathematical expression that captures evolution, and is often described as a mathematical tautology (Price 1970; Price 1972; Frank 1995). There are many derivations from Price's original formula and below I give one just one (El Mouden et al. 2013):

$$\Delta\bar{z} = cov(v, z) + E\text{v}(\Delta z) \tag{3.2}$$

Following El Mouden et al., z denotes a trait value and $\bar{z}$ the mean value of that trait within a population, such that $\Delta\bar{z}$ denotes change in mean trait value within the population from one generation to the next. The term $cov(v, z)$, captures the covariance between values of z and their relative fitness, v. In this case v is calculated by dividing the number of descendants in the next generation, for a given z value, by the population mean number of descendants. In this way $cov(v, z)$ captures natural selection. In Wilson's model we might imagine that selfish individuals tend to reproduce more than altruists. In the subsequent generation there will be more selfish individuals than altruists in the population, and the mean population value of selfishness will increase. Likewise, the mean population value of altruism will decrease. If selfishness is not related in any way to the number of surviving progeny in the next generation, then the population average for selfishness should not change. Again, the same applies for altruism.

$E_V(\Delta z)$ expresses the average change in the value of z across generations, weighted by relative fitness. This element of the equation deals with transmission, or the fidelity of inheritance from one generation to the next. So, if the organisms in question are asexual clonal species, then the next generation will be identical to the first (assuming no mutation) and this expression will equal zero as there is no change in trait value or fitness. It is also the case that this part of the equation deals with fitness related change not caused by selection—so, for example, by drift.

Price has shown that covariance between fitness and a trait value leads to natural selection: indeed this is the formal statistical definition of natural selection (Frank 1995) that emerges from Darwin and was captured by Lewontin's general scheme (Lewontin 1970). It was Price's work that caused Hamilton to more thoroughly develop and formalize his model of inclusive fitness (Hamilton 1970). Okasha claims that the Price Equation enables effective modelling of the interaction of effects at different levels of biological organization and he gives a two-level example—individuals and groups.

> In a "multi-level" evolutionary scenario, where smaller replicating units are nested within larger ones, Price's equation permits us to describe the combined effects of two (or more) levels of selection on the overall evolutionary change. This is because natural selection at one level affects the transmission fidelity at the next level up, thus leading to recursions between levels. So for example, suppose the two levels are individuals and groups. By applying the basic Price equation to the groups, as described above, we can write the overall evolutionary change as the sum of two terms, one reflecting selection on the groups, the other the fidelity with which groups transmit their "character" to the next generation. But the latter depends, in part, on selection at the individual (within-group) level. If there is lots of within-group selection, then a group composed, for example, of individuals of two types A and B in equal proportion, may produce offspring predominantly of type A; the group will thus have a low transmission fidelity. So transmission bias at the group level corresponds to selection at the individual level. Thanks to this principle, Price's technique allows us to write the combined effects of individual and group selection in a single equation (Okasha 2005: 999).

Here Okasha is stating that the final portion of the equation, $Ev(\Delta z)$, that permits fitness related change not due to direct selection can be co-opted to model within-group selection effects, in the context of an overall group level selection model. Thus, the covariance term will apply at the group level, and other changes, including within-group selection caused changes, will complete the transmission, or fidelity term.

This leads Okasha to three conclusions: (i) that the amount of variance at each level of biological organization determines the balance of selection; (ii) that selection at lower levels reduces heritability at higher levels; and (iii) that for higher-level selection to beat the effects of lower-level selection, mechanisms that reduce conflict within lower levels may be necessary. Okasha notes that the emergence of mechanisms of this type have been linked to the major transitions in evolution, enabling the emergence of distinct levels of biological organization including that of the individual (Maynard-Smith and Szathmary 1995; Queller and Strassmann 2009). In all likelihood this makes MLS1 events extremely rare, and gene- or individual-level selection much more probable. Nonetheless, Okasha's adoption of Price gives theoretical reasons to entertain the hypothesis, and this is in keeping with Williams who argued that classic group selection was unlikely due to migration and gene flow effects, but not impossible. Both Okasha and Williams have been keen to apply rigorous standards and the use of the Price Equation perhaps only serves to strengthen Williams' general points about how to think about adaptation and natural selection. To this end, MLS theory models are firmly a part of the emerging adaptationism that Gould has referred to.

3.3.3.1 Criticisms of the Hardened View

Dawkins' replicator-vehicle distinction, derived from Williams and others, expressed the emerging separation of molecular genetics from functional discussions in the latter phase of the Modern Synthesis. This relates directly to Gould's argument that natural selection and adaptationism began to hold prime position, such that phenotypes could be discussed as adaptations which in turn allowed modelling under the assumption that there were single *genes for* single traits, sometimes referred to as the phenotypic gambit (Grafen 1984).[11] Related to this was the idea that genes can be heuristically viewed as strategists with their strategies embodied in vehicular traits. Those strategies will have some impact on fitness and the competitive strategic dynamics of traits across generations could be modelled to see which would stabilize over time, which were evolutionarily more successful (Maynard-Smith 1982).

Strict adherents to selfish gene theory will see models of individual selection as a useful heuristic that works only when properly constrained by the underlying assumptions of gene-level selection that includes an understanding of how genes cooperate to form individuals (West and Gardner 2013).

Selfish gene theory has received much criticism, with an especial focus upon the fundamental assumptions in play (see Godfrey Smith 2010; Okasha 2006). For instance, genes themselves are the products of evolution, with an evolutionary history that relied upon evolutionary processes prior to genes. This kind of argument conflates replicators with the process, and the general response has been to hypothesize about the kinds of early replicators that might have initiated biological evolution as it is currently understood. Similarly, critics note that the biological hierarchies

[11] It should be noted that Wilson's (1975) model was a standard case of this, relying on a haploid population and a single gene that could be in either altruistic or selfish allelic form.

assumed in standard models are also a product of evolution and again require an account. Commonly this kind of claim is made within the context of theoretical work on the major transitions in evolution (Maynard-Smith and Szathmary 1995). So, Dawkins' characterization of replicators depended upon a view of genes as catalysts. This means that genes are foundational for the levels of organization above because catalytic genes and their actions represent necessary conditions. Given this, if genes are the outcome of evolutionary processes, then prior biological organization must also have permitted that evolutionary outcome, and we can therefore hypothesize that evolution is not constrained to the gene level in the present. More strongly, as we have seen, it has been argued that selection must happen at multiple levels in order to explain the major transitions that led to the formation of biological organization (Okasha 2003, 2005). Again, this argument can be met with hypotheses about pre-genetic replicators, but also with a note that MLS1 positions measure fitness at the individual level, and that both MLS1 and MLS2 outcomes are likely very rare.

The arguments summarized above stem from a desire to push back the origins of biology. They assume evolution in the most minimal sense, the sense of changes in form across time, and genes have distinct form. As noted, what such a view misses are the broad abstractions that Williams and then Dawkins furnished, that enable a more generic understanding of how selection dynamics could emerge. Improvements in *copying fidelity*, *fecundity,* and *longevity* in ancestral molecules would facilitate stability in evolutionary transitions. Put more plainly, replicators can be more or less good at replicating, and this will feed into evolutionary transitions.

Other criticisms of selfish gene theory are based upon an informational interpretation of genes. Dawkins, as we have seen, discusses the role of genes in development in catalytic terms. But he also introduces an informational view and tells us that organisms inherit the "*ideas* of ancestral design, in the form of the DNA program" (Dawkins 1989: 261). For Dawkins the notion of a program implies that development is a new process each time it happens, but it is catalyzed by genes within their appropriate contexts. He gave the example of individuals not inheriting the same heart as their mother or father, but instead building their own by using genes inherited from their parents to catalyze the process. Indeed, in this discussion, Dawkins was clear to note that selection can only sift genes as a result of their embedded, phenotypic effects.

Dawkins' language here is readily interpreted as highly deterministic if close attention is not paid to the detail of his argument. *Ideas* are representational, and together with the notion of a *program* he could appear to commit to a concept of information as a complete instructional package. What he meant by DNA program is in fact a stereotyped developmental process that has DNA as an initial and necessary input—DNA is data, that is made sense of within a context. As will be discussed in Chap. 5 it is the functional relationship between data and context that delivers information, and thus DNA has an informational role within the broader developmental context. Neither Dawkins, nor Williams and other predecessors committed to Laplacean determinism.

While gene-level considerations are a clear part of the emerging focus upon adaptation and natural selection, Gould made other criticisms of the hardening of the Modern Synthesis. Most notably he complained about what he saw as a pan-adaptationist approach emerging within biology, in which adaptation was simply assumed with low evidential standards, something which went against the cautions given by Williams but more broadly questioned the notion of what it is for something to be a trait given that individuals were integrated wholes (Gould and Lewontin 1979).

In his paper with Lewontin, Gould noted that much of what may look like an adaptation is there due to developmental constraints leading to restrictions in design opportunity—tolerable oncosts of selected design, perhaps. Later, he suggested the idea that adaptationist hypotheses developed in the context of current function may not reflect the historical functionality of those traits, and that they may have been co-opted, or exapted to modern utility (Gould and Vbra 1982). Broadly speaking these criticisms were based on a view of adaptation as an empirical hypothesis in which natural selection has wide reach and is the prime mover in evolution, a claim Gould saw as unsupportable for the reasons given (Forber 2008b). One response to this has been to regard adaptationism as a methodological approach (Andrews et al. 2002; Dennett 1995). Here adaptationist hypotheses are mooted to give a clear and testable situation. Such hypotheses may be "just-so" stories but if crafted with due diligence, and awareness of Williams' discussion of complexity, constancy, and the onerous nature of supporting such claims, then meaningful work can be done. Williams' discussion of the lateral line work in fish fitted this view. As Forber (2008b) also notes, there is a stronger explanatory claim that apparent design is the main target of evolutionary biology, as a discipline, and natural selection is the principal framework for explaining it (see also (Lewens 2008)). Finally, Gould was also challenging both empirical and explanatory views when noting that there are non-adaptationist hypotheses of form available that account for the fit of a trait, and its apparent optimality (Forber 2008a), indeed optimality may not be the preserve of only adaptations (Potochnik 2008).

3.4 Summary

Darwin redefined evolution, moving it from a transformational account toward a variational one. While personally he became more tolerant of Lamarckian views as he further developed his work, to account for the generation of variation, his core idea of variation, differential success, and inheritance delivering natural selection took hold of numerous scientific imaginations. As I have shown, not all imaginations grasped Darwinian views equally, and the re-emergence of Mendel provoked a lengthy debate between the biometricians and Mendelians. It was this debate that was resolved by the emergence of statistical techniques and population genetics. Biology could eat is gradualist cake and have its particulate inheritance too.

Evolution was now resolutely about the emergence of variation in the gene pool and its selection, although genetic drift and other forces could shift the ground and lead to new trajectories for populations. It was no longer possible to incorporate the inheritance of acquired characteristics, and nor was evolution seen as constrained and directional. Orthogenesis was thus removed as a consideration. But all phenomena in biology were up for synthesis within the new statistical genetics.

All this activity mostly took place between 1900 and the early 1940s. Thereafter, more and more biologists began to focus upon adaptation and the role of natural selection. It is not entirely clear why this happened, but it is possible that the epistemic strategy of posing adaptationist hypotheses was well rewarded with early wins and made clear sense as a starting point. Nonetheless, as late as 1966 Williams was complaining about a lack of clarity around adaptation and natural selection and sought to dispel this, with clear and cautious thinking about it. Just over a decade later, Gould and colleagues were complaining that everyone was an adaptationist and without good reason or data. Gould wanted a return to the pluralism of what he called (perhaps now ironically) the restrictive phase of the Modern Synthesis.

As well as concerns about adaptationism as a research program there were concerns about the conception of the gene that emerged during the hardening phase. Williams again captured this, and Hamilton and others further developed the abstract, non-molecular gene concept within inclusive fitness theory, leading to great gains in the explanation of cooperation. But Dawkins' *Selfish Gene*, with his replicator-vehicle distinction, while a great work of detailed and subtle scholarship, surveying the second part of the Modern Synthesis, adopted metaphors and phrases that perhaps led to the interpretation of the gene-level view as gene-centric and overly deterministic in its relation to form. It is this book that I see as the final act of the development of the Modern Synthesis, because it so clearly captured the field but also developed the ideas to their fullest extent. From 1976 onward biologists applied the Modern Synthesis, but some also began to criticize it in a concerted way.

The role of genes in Modern Synthetic accounts and the role of development as a constraint on form (from Gould and Lewontin 1979) became central concerns of those critically engaging with this second phase of the Modern Synthesis. It is in this first wave of critics—Gould and Lewontin principal among them—that the Extended Synthesis finds its conceptual ancestors. Some also lay claim to multi-level selection theory as a possible extension to the Modern Synthesis and see Wynne-Edwards as at least inspiring new challenges. But as I hoped to show above, this approach really relies upon gene-level accounts to populate at least some levels (in MLS1), as well as standard conceptions of fitness (in MLS1) and its most stringent support comes from use of the Price Equation, which is a mathematical description of natural selection. In this way multi-level selection theory does not rely on other evolutionary processes.

In the following chapters I will draw out key arguments from developmental and anti-gene centric positions and show how they work. At root all the arguments rely upon some kind of theory of information, either implicitly or explicitly, and this presents a pathway to possible resolution.

References

Allen GE (1969) Hugo de Vries and the reception of the "mutation theory". J Hist Biol 2(1):55–87. https://doi.org/10.1007/BF00137268

Andrews PW, Gangestad SW, Matthews D (2002) Adaptationism – how to carry out an exaptationist program. Behav Brain Sci 25(4):489–504. http://www.ncbi.nlm.nih.gov/pubmed/12879701

Axelrod R (1990) The evolution of cooperation. Penguin Books

Axelrod R, Hamilton W (1981) The evolution of cooperation. Science 211:1390–1396. http://www.sciencemag.org/content/211/4489/1390.short

Becquemont D (2011) Social Darwinism: from reality to myth and from myth to reality. Stud Hist Philos Sci C Stud Hist Philos Biol Biomed Sci 42(1):12–19. https://doi.org/10.1016/j.shpsc.2010.11.001

Bowler PJ (2017) Alternatives to Darwinism in the early twentieth century. In: Delisle RG (ed) The Darwinian tradition in context: research programs in evolutionary biology. Springer, New York, pp 195–217. https://doi.org/10.1007/978-3-319-69123-7_9

Cook LM, Saccheri IJ (2013) The peppered moth and industrial melanism: evolution of a natural selection case study. Heredity 110(3):207–212. https://doi.org/10.1038/hdy.2012.92

Coyne JA, Barton NH, Turelli M (1997) Perspective: a critique of Sewall Wright's shifting balance theory of evolution. Evolution 51(3):643–671. https://doi.org/10.1111/j.1558-5646.1997.tb03650.x

Coyne JA, Barton NH, Turelli M (2000) Is Wright's shifting balance process important in evolution? Evolution 54(1):306–317. https://doi.org/10.1111/j.0014-3820.2000.tb00033.x

Damuth J, Heisler ILL (1988) Alternative formulations of multilevel selection. Biol Philos 3 (4):407–430. https://doi.org/10.1007/BF00647962

Dawkins R (1976) The selfish gene. Oxford University Press, Oxford

Dawkins R (1989) The selfish gene, 2nd edn. Oxford University Press

Delisle RG (ed) (2017) The Darwinian tradition in context. Springer Nature Switzerland. https://doi.org/10.1007/978-3-319-69123-7

Delisle RG (ed) (2021) Natural selection: revisiting its explanatory role in evolutionary biology. Springer

Dennett DC (1995) Darwin's dangerous idea: evolution and the meanings of life. The Penguin Press, Allen Lane

Depew DJ (2010) Darwinian controversies: an historiographical recounting. Science and Education (Vol. 19, Issues 4–5). https://doi.org/10.1007/s11191-009-9202-x

Depew DJ (2017) Natural selection, adaptation, and the recovery of development. In: Huneman P, Walsh DM (eds) Challenging the modern synthesis: adaptation, development, and inheritance. Oxford University Press, Oxford, pp 37–67

Depew DJ, Weber BH (1996) Darwinism evolving: systems dynamics and the geneaology of natural selection. MIT Press

Dickins TE, Rahman Q (2012) The extended evolutionary synthesis and the role of soft inheritance in evolution. Proc R Soc B Biol Sci 279(1740):2913–2921. https://doi.org/10.1098/rspb.2012.0273

Dietrich MR, Skipper RA Jr (2012) A shifting terrain: a brief history of the adaptive landscape. In: Svensson EI, Calsbeek R (eds) The adaptive landscape in evolutionary biology. Oxford University Press, Oxford, pp 3–15

Edwards AWF (2008) G. H. Hardy (1908) and Hardy-Weinberg equilibrium. Genetics 179 (3):1143–1150. https://doi.org/10.1534/genetics.104.92940

El Mouden C, André J-B, Morin O, Nettle D (2013) Cultural transmission and the evolution of human behaviour: a general approach based on the Price equation. J Evol Biol, 1–11. https://doi.org/10.1111/jeb.12296

Esposito M (2011) Utopianism in the British evolutionary synthesis. Stud Hist Philos Sci C Stud Hist Philos Biol Biomed Sci 42(1):40–49. https://doi.org/10.1016/j.shpsc.2010.11.007

Fisher RA (1918) The correlation between relatives on the supposition of Mendelian inheritance. Trans R Soc Edinburgh 52(2):399–433. https://doi.org/10.1017/S0080456800012163
Fisher RA (1922) On the dominance ratio. Trans R Soc Edinburgh 42:321–341
Fisher RA (1930) The genetical theory of natural selection. Clarendon Press
Fisher RA (1936) Has Mendel's work been rediscovered? Ann Sci 1:115–137. https://doi.org/10.1080/00033793600200111
Forber P (2008a) Spandrels and a pervasive problem of evidence. Biol Philos 24(2):247–266. https://doi.org/10.1007/s10539-008-9144-8
Forber P (2008b) Introduction: a primer on adaptationism. Biol Philos 24(2):155–159. https://doi.org/10.1007/s10539-008-9150-x
Frank SA (1995) George Price's contributions to evolutionary genetics. J Theor Biol 175 (3):373–388. https://doi.org/10.1006/jtbi.1995.0148
Frank SA (1998) Foundations of social evolution. Princeton University Press
Frank SA, Slatkin M (1992) Fisher's fundamental theorem of natural selection. Trends Ecol Evol 7 (3):92–95. https://doi.org/10.1016/0169-5347(92)90248-A
Futuyma DJ (2015) Can modern evolutionary theory explain macroevolution? In Macroevolution: interdisciplinary evolution research (Vol. 2, pp. 29–86). Springer. https://doi.org/10.1007/978-3-319-15045-1
Gavrilets S (1997) Evolution and speciation on holey adaptive landscapes. Trends Ecol Evol 12 (8):307–312. https://doi.org/10.1016/S0169-5347(97)01098-7
Godfrey Smith P (2010) Darwinian populations and natural selection. Oxford University Press
Gould SJ (1977) Ontogeny and phylogeny. The Belknap Press of Harvard University Press
Gould SJ (2002) The structure of evolutionary theory. The Belknap Press of Harvard University Press
Gould SJ, Lewontin RC (1979) The spandrels of San Marco and the Panglossian paradigm: a critique of the adaptationist programme. Proc R Soc Lond Biol Sci 205(1161):581–598. https://doi.org/10.1098/rspb.1979.0086
Gould SJ, Vbra ES (1982) Exaptation – a missing word in the science of form. Paleobiology (Vol. 8, pp. 4–15)
Grafen A (1984) Natural selection, kin selection and group selection. In: Krebs JR, Davies NB (eds) Behavioural ecology, 2nd edn. Blackwell Publishers, New York, pp 62–84
Haig D (2007) Weismann Rules! OK? Epigenetics and the Lamarckian temptation. Biol Philos 22 (3):415–428. https://doi.org/10.1007/s10539-006-9033-y
Hamilton WD (1964) The genetical evolution of social behaviour. I. J Theor Biol 7(1):1–16. https://doi.org/10.1016/0022-5193(64)90038-4
Hamilton WD (1970) Selfish and spiteful behaviour in an evolutionary model. Nature 228 (5277):1218–1220. https://doi.org/10.1038/2281218a0
Harman O (2010) The price of altruism: george price and the search for the origins of kindness. W.W. Norton, New York
Huxley JS (2010) Evolution: the modern synthesis: the definitive edition. MIT Press
Joshi A (1999) The shifting balance theory of evolution. Resonance 4(12):66–75. https://doi.org/10.1007/BF02838675
Kauffman SA (1993) The origins of order: self-organization and selection in evolution. Oxford University Press, Oxford
Koonin EV, Wolf YI (2009) Is evolution Darwinian or/and Lamarckian? Biol Direct 4:42. https://doi.org/10.1186/1745-6150-4-42
Kutschera U, Niklas K (2004) The modern theory of biological evolution: an expanded synthesis. Naturwissenschaften 91(6):255–276. https://doi.org/10.1007/s00114-004-0515-y
Lewens T (2008) Seven types of adaptationism. Biol Philos 24(2):161–182. https://doi.org/10.1007/s10539-008-9145-7
Lewontin RC (1970) The units of selection. Annu Rev Ecol Syst 1(1970):1–18. http://www.jstor.org/stable/10.2307/2096764

Loison L, Herring E (2017) Lamarckian research programs in French biology (1900–1970). In: Delisle RG (ed) The Darwinian tradition in context: research programs in evolutionary biology. Springer, New York, pp 243–269. https://doi.org/10.1007/978-3-319-69123-7_11
Maynard Smith J (1976) Group selection. Quaterly review of biology (Vol. 51, Issue 2, pp. 277–283)
Maynard-Smith J (1982) Evolution and the theory of games. Cambridge University Press, Cambridge
Maynard-Smith J, Szathmary E (1995) The major transitions in evolution. Oxford University Press, Oxford
Mayr E (1982) The growth of biological thought: diversity, evolution and inheritance. Belknap Press
Mayr E (1993) What was the evolutionary synthesis? Trends Ecol Evol 8(1):31–34. https://doi.org/10.1016/0169-5347(93)90128-C
Okasha S (2003) Original article recent work on the levels of selection problem by Samir Okasha. Hum Nat 3:349–356
Okasha S (2005) Maynard Smith on the levels of selection question. Biol Philos 20:989–1010. https://doi.org/10.1007/s10539-005-9019-1
Okasha S (2006) Evolution and the levels of selection. Oxford University Press, Oxford
Oyama S (2000) The ontogeny of information: developmental systems and evolution, 2nd edn. Duke University Press
Pigliucci M (2007) Do we need an extended evolutionary synthesis? Evol Int J Org Evol 61 (12):2743–2749. https://doi.org/10.1111/j.1558-5646.2007.00246.x
Potochnik A (2008) Optimality modeling in a suboptimal world. Biol Philos 24(2):183–197. https://doi.org/10.1007/s10539-008-9143-9
Price GR (1970) Selection and covariance. Nature 227:520–521
Price GR (1972) Extension of covariance selection mathematics. Ann Hum Genet 35(4):485–490. https://doi.org/10.1111/j.1469-1809.1957.tb01874.x
Provine WB (2001) The origins of theoretical population genetics, 2nd edn. University of Chicago Press
Queller DC, Strassmann JE (2009) Beyond society: the evolution of organismality. Philos Trans R Soc Lond Ser B Biol Sci 364(1533):3143–3155. https://doi.org/10.1098/rstb.2009.0095
Reif WE, Junker T, Hossfeld U (2000) The synthetic theory of evolution: general problems and the German contribution to the synthesis. Theory Biosci 119(1):41–91. https://doi.org/10.1007/s12064-000-0004-6
Romanes GJ (1888) Lamarckism versus Darwinism. Nature 38:413
Scott-Phillips T, Dickins TE, West SA (2011) Evolutionary theory and the ultimate–proximate distinction in the human behavioral sciences. Perspect Psychol Sci 6(1):38–47. https://doi.org/10.1177/1745691610393528
Skipper RA Jr, Dietrich MR (2012) Sewall Wright's adaptive landscape: philosophical reflections on heuristic value. In: Svensson EI, Calsbeek R (eds) The adaptive landscape in evolutionary biology. Oxford University Press, Oxford
Smocovitis VB (1992) Unifying biology: the evolutionary synthesis and evolutionary biology. J Hist Biol 25(1):1–65. https://doi.org/10.1007/BF01947504
Smocovitis VB (1996) Unifying biology: the evolutionary synthesis and evolutionary biology. Princeton University Press
Sober E, Wilson DS (2011) Adaptation and Natural Selection revisited. J Evol Biol 24(2):462–468. https://doi.org/10.1111/j.1420-9101.2010.02162.x
Svensson EI (2016) Adaptive landscapes. Encycl Evol Biol, November, 9–15. https://doi.org/10.1016/B978-0-12-800049-6.00054-8
Trivers R (1971) The evolution of reciprocal altruism. Q Rev Biol 46(1):35–57. http://www.jstor.org/stable/10.2307/2822435

Ulett MA (2014) Making the case for orthogenesis: the popularization of definitely directed evolution (1890–1926). Stud Hist Philos Sci C Stud Hist Philos Biol Biomed Sci 45 (1):124–132. https://doi.org/10.1016/j.shpsc.2013.11.009
Vorzimmer P (1963) Charles Darwin and blending inheritance. Isis 54(3):371–390. https://doi.org/10.1086/349734
Webb RH (2011) If evolution is the answer, what is the question? J Evol Psychol 9(2):91–107. https://doi.org/10.1556/JEP.9.2011.2.2
Weismann A (1889) Essays upon Heredity and Kindred Biological Problems. (E. Poulton, S. Schonland, & A. E. Shipley (Eds.); Authorized). Clarendon Press. https://www.gutenberg.org/files/48132/48132-h/48132-h.htm
Weismann A (1893) The all-sufficiency of natural selection: a reply to Herbert Spencer. Contemp Rev 64:309–338
West SA, Gardner A (2013) Adaptation and inclusive fitness. Curr Biol 23(13):R577–R584. https://doi.org/10.1016/j.cub.2013.05.031
West SA, Griffin AS, Gardner A (2007) Social semantics: altruism, cooperation, mutualism, strong reciprocity and group selection. J Evol Biol 20(2):415–432. https://doi.org/10.1111/j.1420-9101.2006.01258.x
Williams GC (1996) Adaptation and natual selection: a critique of some current evolutionary thought, 2nd edn. Princeton University Press
Wilson DS (1975) A theory of group selection. Proc Natl Acad Sci USA 72(1):143–146. https://doi.org/10.1073/pnas.72.1.143
Wilson DS (2008) Social semantics: toward a genuine pluralism in the study of social behaviour. J Evol Biol 21(1):368–373. https://doi.org/10.1111/j.1420-9101.2007.01396.x
Wright S (1931) Evolution in Mendelian populations. Genetics 16(March):97–158. https://doi.org/10.1007/BF02459575
Wright S (1932) The roles of mutation, inbreeding, crossbreeding and selection in evolution. In Proceedings of the sixth international congress of genetics (pp. 356–366)
Wright S (1988) Surfaces of selective value revisited. Am Nat 131(1):115–123
Wynne-Edwards VC (1962) Animal dispersion in relation to social behaviour. Oliver and Boyd

4 Causation

Abstract

The Modern Synthesis can be regarded as an attempt to unify biology and to protect it from reduction to chemistry and physics, and thus to preserve the identity of biology as a discipline. Mayr was particularly sensitive to this aspect of the synthesis and developed a specific account of biological causation in part to separate biology from other disciplines. He published his case in 1961, making a distinction between proximate and ultimate causation. This distinction and Mayr's model of causation have been heavily criticized by advocates of an Extended Evolutionary Synthesis. In this chapter, I detail Mayr's original argument and then core arguments from those opposing his view. I defend Mayr analytically, but I also make a comment on the possibility that Mayr and his critics are simply operating different forms of idealization to deliver on different tasks. If this is the case, I suggest, then Mayr's view has not really been dismissed as false but rather positioned within specific task demands.

Keywords

Unification · Positivism · Causation · Proximate · Ultimate · Reciprocal causation · Idealization · Niche construction

4.1 Introduction

Smocovitis places the evolutionary synthesis, which she sees as a historically more sensitive term than the Modern Synthesis (personal communication) within a broader, positivist project of unifying biology (Smocovitis 1992, 1996). A basic ambition of positivism, and the logical positivism emerging from the Vienna Circle of the 1920s and 1930s, was to remove all metaphysical trappings from science and to ground it instead in axiomatic principles. In this way, all of the sciences might become unified, and Smocovitis details the history of calls to axiomatize biology

T. E. Dickins, *The Modern Synthesis*, Evolutionary Biology – New Perspectives on Its Development 4, https://doi.org/10.1007/978-3-030-86422-4_4

from Woodger, Haldane, Julian Huxley and others. But she also points to a tension or fear that an axiomatic biology might be reduced to chemistry and physics and lose its identity in a generic unification. It was the emergence of an evolutionary theory cleansed of metaphysical assumptions, during the synthesis, that was heralded as a source of unification within biology, but also as the mark of its independence as a discipline.

Smocovitis sees evolution, prior to the synthesis, as a discipline that reeked of "metaphysical and vitalistic elements—witness the popularity of such views as directed evolution, Lamarckism, emergent evolution, and the numerous other agents of evolution outlined by Darwin and his heirs—but it also defied the great method for grounding positivistic claims to knowledge: experimentation" (1992, 14). The early twentieth century rise of genetics gave both mechanism and experiment and generated lawlike statements. This did not initially help evolutionary studies that were still grounded in natural, historical methods of observation, and they found themselves in decline within universities in want of funding. According to Smocovitis, this was another cause of the eclipse of Darwin (Sect. 3.2). For Smocovitis, the emergence of experimentation, and in particular mathematical modelling that lent objectivity to experimental outcomes, brought evolution into the empirical fold and began the process of unification. This explains why Fisher, Haldane, and Wright were seen as such influential actors in the synthesis (Sect. 3.3).

> With the work of the modelers, and the adoption of experimentation, the view of natural selection as agent began to diminish, so that by the 1930s natural selection took on a causo-mechanical existence. Terms borrowed from the physical sciences, like "cause" (Haldane's preferred word), "factor" (Wright's preferred word), and finally "mechanism" (Dobzhansky's and Huxley's preferred word) slowly supplanted the term and the view of selection as agent, although Huxley viewed selection as both agent and mechanism simultaneously and Fisher still viewed it as an agent. (Smocovitis 1992: 20)

Natural selection was now becoming something that could be both measured and tested, and importantly a part of a physical, mechanistic world view. Conceptions of the gene followed suit such that the synthesis can be seen as the "bringing together of the material basis of evolution (the gene) with the mechanical cause of evolutionary change (selection), to make a mechanistic and materialistic science of evolution that rivaled Newtonian physics while still preserving the autonomy of the biological sciences" (Smocovitis 1992: 24). Gould's phase of restriction occurred as this unification removed the more metaphysical commitments from evolution, such as Lamarckism and orthogenesis as directed evolution (Gould 2002). Smocovitis (1992) also credits Julian Huxley with pushing natural selection as a principal source of evolutionary change in his 1942 book, which she analyzes as an attempt to marry unified biology with the progressive ideals of the Enlightenment, albeit in a world enmired in the horrors of World War II. This was Gould's hardening phase.

The unification of biology, caused by the synthesis, was a major achievement and placed biology firmly within the empirical and axiomatic sciences. However, genetics and the increasing sophistication of molecular biology through the 1950s again raised the threat of biology being completely reduced to chemistry and losing its identity. Smocovitis (1992) argues that Mayr was the most philosophically astute among the founders of the synthesis and he invested in defending biology from this fate. A crucial part of Mayr's defensive work was his effort to separate biological causation from that found in physics, and it is this topic that the remainder of this chapter addresses. Mayr's statement on causality can be read as the principal philosophical statement of the Modern Synthesis, and unsurprisingly it has attracted the critical attention of those advocating for an extension. I will first turn to Mayr's classic paper on causation and then to its critics.

4.2 Causation

In 1961 Ernst Mayr published *Cause and Effect in Biology* in which he discussed the kinds of causes available to biological science, as well as views on predictability, teleology, and probability (Mayr 1961). Most famously, this paper clarified the distinction between proximate and ultimate causes, which has been widely adopted as a framework for organizing biological accounts. As we shall see, however, this distinction has come under considerable critical scrutiny in the twenty-first century, with some theorists advocating that it should be abandoned in favor of reciprocal causation (Laland et al. 2011).

Mayr's paper can be regarded as a foundational document of the Modern Synthesis in that it takes the reader through a set of philosophical considerations about causation and biological explanation that characterize the frame of thinking developed in the Modern Synthesis. Mayr's is not a paper about the specifics of any particular evolutionary account but rather about what is expected of any such argument. In this light, recent criticisms of the proximate–ultimate distinction can be seen as analyses of the epistemological core of the Modern Synthesis. Elimination of the proximate–ultimate distinction and its replacement with a different model of causation could potentially amount to a shift of perspective away from the Modern Synthesis.

4.2.1 Proximate and Ultimate Causation

Mayr opens his paper apologetically, noting that he feels unable to deliver a logician's analysis of causality but that he hopes to outline the issue of causation that is relevant to biologists. He then makes the first of a number of comments about vitalism, placing that position in opposition to materialism, before quickly outlining the core components of causality which "is believed to contain three elements: (i) an explanation of past events ("a posteriori causality"), (ii) prediction of future events; (iii) interpretation of teleological—that is, "goal-directed"—phenomena" (1961: 1501). This tripartite view he attributes to Ernest Nagel and this informs the structure of much of the second half of the paper. Before he turns to that analysis Mayr first seeks to clarify just what biology, as a discipline, is focused upon.

For Mayr, biology is comprised of two subdisciplines that differ in methods, in the framing of central problems, and in core concepts. The first subdiscipline is *functional biology*, with a focus on the mechanisms at work at all levels of the biological hierarchy. This workstream is characterized by the *how* question, which incorporates how something works or how it functions, relying upon experimental methods that enable isolation of key variables and the control of others. The term, *functional biology,* can cause confusion in contemporary circles more used to applying functional analyses separately from mechanistic ones, but in this case, Mayr means mechanistic analysis—the function may be understood, but it is the *how* of its delivery which is the focus here.

The second subdiscipline is that of *evolutionary biology*, in which biologists ask and attempt to answer the historical question of how something has come to be. This can be packaged as a form of *why* question such that one can ask why any given trait exists. Mayr is very clear that few biological structures can be fully understood unless inspected against this long historical background of evolution, and he emphasizes this point by noting that any focal organism is simply a point in time and space in an ongoing dynamic system.[1]

Once this disciplinary subdivision is made, Mayr reinterprets it in terms of information:

> The functional biologist deals with all aspects of the decoding of the programmed information contained in the DNA code of the fertilized zygote. The evolutionary biologist, on the other hand, is interested in the history of these codes of information and in the laws that control the changes of these codes from generation to generation. In other words, he is interested in the causes of these changes. (Mayr 1961: 1502)

Mayr immediately follows this statement with an important qualification. The codes he refers to are not to be seen as inflexible, and he advances examples such as learning and memory to demonstrate that their code structures are what he refers to as open. In modern parlance, we might refer to them as enabling plasticity. That openness is not without restriction, such that processes like learning have degrees of freedom and development conforms to standard views on reaction norms. Thus, learning can differ across the lifespan in terms of what kinds or types of associations are learnable and learned, but there is some freedom in terms of the tokens of that learning. What Mayr is referencing here is the idea of constrained plasticity, and he places his learning example within a discussion of two modes of imprinting in birds, one where birds have a ready representation of what to follow and the other where they will imprint upon the first organism they encounter.

[1] This point is related to some recent discussions about process ontology, and the idea of temporary stability, within the context of discussions about extending the synthesis (Dupre 2017). Dupre also argues that scientists cannot help but adhere to some metaphysics, even if they are naturalistic, or mechanistic. Smocovitis might claim, then, that Mayr's efforts to defend the individuality of biology caused him to embrace some metaphysical principles.

The core question for Mayr is whether causes are the same in functional and evolutionary biology, and he addresses this by producing four plausible causal accounts of warbler[2] migration:

1. *Ecological cause*: as insectivores, warblers must leave an area when winter comes because this suppresses invertebrate populations and will lead to starvation.
2. *Genetic cause*: evolutionary processes have given the warbler the requisite genotype that delivers a phenotype that is sensitive to relevant cues from the environment.
3. *Intrinsic physiological cause*: the warbler responds to decreases in daylight hours and once below a key threshold it leaves and heads to warmer climes.
4. *Extrinsic physiological cause*: a specific and sudden drop in temperature triggered an already prepared bird to leave at precisely that point.

These four causes can be grouped into two. Both of the physiological causes are what Mayr refers to as *proximate causes*, the causes of functional biology. In this case, they capture the bird responding to key environmental factors leading to readiness for migration. The ecological and genetic causes are what Mayr terms *ultimate causes* and are associated with the evolutionary process that give rise to the genetic code within the warbler. This then is the business of evolutionary biology.

Mayr was not the first to use the proximate–ultimate distinction. Lack adopted the terminology in 1950, attributing it to Baker's discussion of the breeding biology of birds (Lack 1950). Accordingly, for Baker, ultimate factors were those that operated via natural selection to bring birds' breeding seasons into alignment with peak prey abundance, whilst proximate factors delivered birds ready to breed at the appropriate time via physiological processes operating in the gonads. As with Baker and Lack, Mayr's claim is that the correlation between migration and available food resources is achieved via natural selection. The individual birds switch to migratory behavior due to physiological responses to light and temperature, those responses being the outcome of DNA building the relevant phenotype. Mayr summarizes the idea as follows:

> Still another way to express these differences would be to say that proximate causes govern the responses of the individual (and his organs) to immediate factors of the environment while ultimate causes are responsible for the evolution of the particular DNA code of information with which every individual of every species is endowed. (Mayr 1961: 1503)

None of the explanations given for warbler migration are exclusive of one another. Indeed, Mayr's ecological cause provides basic survival pressures to drive selection on relevant genes that will lead to migration and successful food tracking. Those genes will exhibit phenotypic effects through at least the two proximate physiological mechanisms suggested, and it is these effects that selection will have attended to. In this way, it is clear that Mayr meant proximate and ultimate to be seen as

[2]Mayr does not give a species.

complementary causal accounts of any given trait, but he must also have been aware of the interaction between proximate action and selection over historical time given his view of evolutionary biology. Nonetheless, he saw the business of doing functional and evolutionary biology, the practice of those subdisciplines, as logically separable and argued that this was how things proceeded. Indeed, as he wryly alludes in his paper, this practice can often lead to needless debates between biologists, as they talk past one another each asserting their own causal account of some phenomenon to be the right one.

Mayr closes his discussion of proximate and ultimate causation by noting another view on causation from philosophy. A cause can be seen as a "non-sufficient condition, without which an event would not have happened" or as one of a set of "jointly sufficient reasons without which the event could not have happened" (1961: 1503). The strengths of this view are evident, for Mayr, in simple accounts of proximate mechanisms, but they are less evident when dealing with ultimate accounts of complex systems.

4.2.2 Teleology

Aristotle's final cause is described in the context of development, which is described as an orderly and purposive process leading from the egg to the final adult form. It is the final cause that is responsible for this kind of process, which Mayr terms goal-seeking and hence open to teleological accounts. He makes quick work of pointing out error-prone accounts of purposive processes, again alighting upon a version of vitalism and labelling it supernatural in flavor.[3] What Mayr wants is an account of purposiveness set in purely materialist terms, and he asks when it is appropriate to label something purposive without falling foul of vitalism, etc. His answer is that an "individual who—to use the language of the computer—has been "programmed" can act purposively. Historical processes, however, can *not* act purposively" (1961: 1503).

Mayr's migrating warbler is programmed to migrate in response to relevant environmental cues, and the code is inherent or instantiated in the DNA. Mayr argues that the DNA code that "controls the development of the central and peripheral nervous systems, of the sense organs, of the hormones, of physiology and morphology, is the program for the behavior computer of this individual" (1961: 1504). These programs are the product of natural selection, and again, Mayr makes clear their open nature, which allows for constrained plasticity and for reaction norms. It is this property that enables feedback between the organism (its programs) and the environment to calibrate better and increase survival chances. Mayr therefore directly relates a notion of plasticity of response to fitness maximization. He terms this a purely mechanistic purposiveness.

[3] I take it that this is Mayr dismissively chasing out vestiges of metaphysics.

The purposiveness of behavior or development is the decoding of the program, and this process tests the code in Mayr's scheme. The DNA code that constitutes the program is perfected over evolutionary time by natural selection, which can, of course, lead to radical change over a long time. Mayr cautions us to see these as distinct, with evolution having no goal and the programmed goals of behavior and development as teleonomic rather than teleological; this last term clearly distinguishing systems that have goals as a part of their processes. By implication, those systems require an explanation of their design, and from the preceding arguments, it is clear that this is what evolutionary biology delivers. Again, Mayr shows the complementarity between proximate and ultimate concerns.

4.2.3 Prediction

Mayr opens his discussion of prediction by separating explanation from prediction within evolutionary biology, following Scriven's distinction (Scriven 1959). Evolutionary biology can explain the presence of a trait in historical terms based upon natural selection, but it is not able to predict the outcomes of evolution, the future forms, etc. that will emerge. Mayr not only notes the difficulty or impossibility of evolutionary prediction but more generally shows that prediction in functional biology is problematic and where it works, it is statistical or probabilistic by nature.

> As is true in many other branches of science, the validity of predictions for biological phenomena (except for a few chemical or physical unit processes) is nearly always statistical. We can predict with high accuracy that slightly more than 500 of the next 1000 newborns will be boys. We cannot predict the sex of a particular unborn child. (Mayr 1961: 1505)

Probabilistic explanation is a function of indeterminacy, and Mayr discusses four sources of this:

1. *Randomness of the event with respect to the significance of the event.* Here Mayr gives as an example the now familiar observation that mutation of DNA code is random with respect to the evolutionary outcome (Monod 1971). In this discussion, Mayr also characterizes the process of the "determination of the genetic contents of the zygote" (1961: 1505) as containing much randomness and making it hard to predict the final phenotypic outcome with respect to selection.
2. *Uniqueness of all entities at the higher levels of biological integration.* This uniqueness is peculiar to biology. Mayr notes that in physics successful generalizations can be drawn at a lawlike level, and he gives the example of ice floating on water. A statement that ice will do this is universally applicable for all ice and all bodies of water. Statements of this sort for a biological individual only have statistical currency, although Mayr notes that taxonomic distinctions are more like those of physics. This again makes prediction difficult.
3. *Extreme complexity.* Mayr again borrows from physics and notes the *n*-dimensionality of organic beings, built upon multiple pathways of action,

feedback, and homeostatic mechanism, all of which defeats the project of a complete explanation and full prediction. Again, Mayr is reflecting upon processes of growth, development, and maintenance in these comments.

4. *Emergence of new qualities at higher levels of integration.* Mayr simply states that emergent properties do occur and cannot be predicted, and he cautions that this kind of indeterminacy, as with the preceding three, does not imply a lack of cause merely unpredictability.

All of these sources of indeterminacy reduce our ability to predict, Mayr tells us, but he then relates these observations to the gas laws that inspired Fisher (Sect. 3.3.2.1):

> One may raise the question at this point whether predictability in classical mechanics and unpredictability in biology are due to a difference of degree or of kind. There is much to suggest that the difference is, in considerable part, merely a matter of degree. Classical mechanics is, so to speak, at one end of a continuous spectrum, and biology is at the other. Let us take the classical example of the gas laws. Essentially they are only statistically true, but the population of molecules in a gas obeying the gas laws is so enormous that the actions of individual molecules become integrated into a predictable—one might say "absolute"—result. Samples of five or 20 molecules would show definite individuality. The difference in the size of the studied "populations" certainly contributes to the difference between the physical sciences and biology. (Mayr 1961: 1506)

4.2.4 Lessons

Mayr concludes his paper by drawing out four lessons. The first is that causality in biology is not like that of classical mechanics, although his gas law example is designed to show more modern physics has incorporated classical mechanics as a simpler and special case of causality. His second and third lessons are closely related. The emphasis is upon multiple causes for many biological phenomena, and he makes a case for bracketing kinds of causes into what must be a supervening explanatory framework where the detail within each bracket may or may not be discovered. These multiple causes make predictability difficult and highly probabilistic at best. Mayr is essentially implying, following his interpretation of Scriven's comments, that partial explanations for a phenomenon can be generated within a bracket or subset of brackets, but a complete account is unlikely, and because of the dimensionality and opportunities for randomness at all points, prediction is limited and statistical. Explanation and prediction are two separable tasks.

Mayr's final lesson is to reemphasize the teleonomic purposiveness of organisms, which he regards as a property conveyed by DNA codes,[4] but to deny any teleology in the evolutionary process.

[4]Here Mayr is committing to a form of design thinking.

4.3 Causation Revisited

Fifty years after Mayr's causation paper (Mayr 1961), a group of scientists and philosophers published a critical reappraisal of its central claims (Laland et al. 2011) in the same journal, *Science*. They explored the proximate–ultimate distinction and how Mayr had developed it and then positioned it against the backdrop of various debates about the adequacy of the Modern Synthesis for twenty-first century biological science. Where Mayr's paper can be considered a foundational document for the Modern Synthesis, so Laland et al.'s paper can be considered an attempt to lay the epistemological foundations of an Extended Evolutionary Synthesis. At the time of writing, it is too soon to afford it quite the same status as Mayr's paper, but it has nonetheless had an impact upon theoretical thinking within the evolutionary sciences.[5] In the following sections, I shall relay the arguments of Laland and colleagues and interpret them in light of the preceding discussion of the original Mayr paper.

4.3.1 Limitations of the Proximate–Ultimate Distinction

After opening with a brief description of the impact of Mayr's paper and the fundamental distinction between proximate and ultimate causes, Laland et al. turn to the problems that they perceive with this distinction. Their first claim is that Mayr's example of warbler migration was suitably simple to clearly make the point about proximate and ultimate causation. Four aspects of the example contributed to this simplicity:

1. Migration is clearly an evolved costly behavior.
2. Birds emerged from flightless dinosaurs suggesting that migration is not a primitive condition of avifauna.
3. Migration responds to separate and autonomous features of the external environment and those external features are not affected by migration.
4. The physiology of migration can be understood without reference to the selection of migration and vice versa.

By implication, the authors are claiming Mayr had something like this in mind when he chose his example, and they qualify point 4 by noting that "this seemingly helps justify Mayr's ... stance that researchers could understand evolution without understanding development" (Laland et al. 2011: 1512). It is not immediately clear why the separation of physiological explanations from evolutionary accounts leads

[5] A Google Scholar search conducted on 28 January 2021 gave Laland et al.'s (2011) paper 528 citations, compared to Mayr's (1961) paper with 2186. We should note that anyone citing the former is highly likely to cite the latter also.

to a position on development. To understand this claim, we must look at the two publications by Mayr that the authors cite in support (Mayr 1984, 1992).

In the earlier article, which is essentially a book review,[6] Mayr begins with a brief history of evolution and the Modern Synthesis (Mayr 1984). Late on in the review, Mayr comes to a claim made by Ho and Saunders that development and what they term the nature of the organism are problems that neo-Darwinism cannot make sense of. It is here that Mayr states that it is not the task of Darwinians to explain development, and he references the proximate–ultimate distinction and his views on DNA codes and their decoding during ontogeny. This statement does lead to the conclusion that evolutionary and developmental explanatory tasks are separable, not least because Mayr is positioning development as a proximate level process (or set of processes). But Mayr makes this point in response to claims from some developmental embryologists that evolution *is* a developmental process. He is keen to emphasize that development is a complex situation in keeping with his views on biological complexity and higher-level integration (Sect. 4.2.3) but that it is a different process than those at work in evolution.

In the second cited work, Mayr makes the following remark:

> After about 1970, particularly in connection with the punctuation controversy, more and more often the claim was made that the Synthesis was obsolete, that we needed a new paradigm… The most vocal of the critics were the developmental biologists… The developmental biologists complained that the Synthesis had ignored them. This was really a rather silly claim, because in the 1940s anybody was welcome to join the Synthesis, but the developmental biologists, almost to the last one, vigorously opposed Darwinism and upheld the ideal of transformational evolution. It was they who refused to join the Synthesis rather than the Synthesis keeping them out. (Mayr 1992: 27–28)

Mayr then goes on to examine the claims of the developmental biologists at this time and notes that they did not make a distinction between *proximate* and *ultimate* causes, that they adhered to a form of teleology, that they did not understand evolution as a population-level concept and that some, notably Waddington who is marked out as a more sympathetic developmentalist, were typological (or essentialist) in their thinking. This last is an aspect of transformational evolution, where individuals within a population develop toward greater complexity, and it is thus a Lamarckian idea (Chap. 2). Here Mayr is making a sweeping criticism of the understanding of evolutionary biology within the developmental community while making clear that it was they who chose not to engage with the emergence of the Modern Synthesis. In effect, Mayr is stating that developmental biologists, during this period of the twentieth century, failed to take up the opportunity to integrate their discipline with evolutionary theory.

This last commentary from Mayr affords a vital perspective. Mayr certainly sees *proximate* and *ultimate* explanations as complementary and a part of a full discipline

---

[6] Ho M-W, Saunders PT (1984) Beyond Neo-Darwinism: An introduction to the new evolutionary paradigm. Academic Press.

of biology, but also as activities that are logically separable and lead to separable work. Moreover, development is a proximate set of processes in which the DNA code is decoded within a complex context that adds to and shapes the final form—he has been clear on that in his 1961 paper, especially when considering sources of randomness and indeterminacy in biology. To this extent, the authors of the 2011 analysis are correct. But what they fail to note is that Mayr's real criticism is of the fundamental assumptions of some developmental biologists, notably their adherence to pre-Darwinian ideas of transformation and essentialism. The deeper philosophical argument against these positions, which Mayr feels was dealt with by the emergence of Darwinism (Mayr 1982), makes those developmental biologists wrong also about development. The mere separation of developmental and evolutionary explanatory tasks from one another is an outcome of a more profound rethinking of what development in fact is. To this end, Mayr has directly engaged with development as an evolutionist in order to make clear that it is not an evolutionary process in its own right, as some had thought because it does not unfold at the population level but rather within individuals.

In a later paper, Laland and colleagues (Laland et al. 2013) make an *ad hominem* claim that all of the published historical analysis of Mayr "universally acknowledges Mayr's insistence of the irrelevance of development to evolution" (2013: 796) and cite five publications in which this argument is apparently endorsed. They are particularly keen to stress that Mayr never gave any indication of how to relate development to evolution in a way that did not violate the proximate–ultimate distinction. But as I have shown above, and Barton and I have elsewhere (Dickins and Barton 2012),[7] Mayr was clear that development had delimited plasticity (reaction norms) as a function of natural selection, and thereby saw a clear relationship. His argument was against the conflation of individual and population-level processes and also against specific typological and transformational views that were held by particular developmental biologists in the mid-twentieth century.

The emphasis upon Mayr's simple example allows Laland et al. to claim that many relevant phenomena are more complex and defeat the proximate–ultimate explanatory framework. They choose as their example intersexual selection and tell us that when "a trait evolves through intersexual selection, the source of selection is itself an evolving character"(Laland et al. 2011: 1512) because it is the peahens' preferences that influence the peacocks' tails. They then state that the "ultimate explanation for the male trait is the prior existence of the female preference, proximately manifest in peahen mate choice decisions, shaped by inherited tendencies and modified through experience by development" (Laland et al. 2011: 1512). They add that the ultimate explanation for peahen preferences rests on the available variation in male tails and its relation to fitness. This is what they dub a *reciprocal* process—or reciprocal causation—in which the proximate mechanisms both respond to and shape natural selection and the authors claim various consequences of this dialectical view (Chap. 8).

[7] It was this paper to which Laland et al. (2013) were reacting.

Before moving to those consequences, it is worth inspecting the detail of this example. First, there is an error that becomes rather conspicuous on re-reading Mayr's discussion of developmental biology (Mayr 1992). Laland et al. refer to traits evolving and evolving characters. Their emphasis is upon the phenotype, and they fail to position evolution as a population level concept. Traits do not evolve, populations do. This suggests some switching between individual and population levels which might well make Mayr's distinction seem less relevant.[8]

Second, their example explicitly relies upon the terms *proximate* and *ultimate* to do its work while subtly conflating their meaning. This conflation takes the form of a gallinovular problem[9] in that the authors decide to start their account from a particular point in time, the existing bias in female preference. This preference continues to drive mate choice, and those choices are between highly heritable and fitness relevant trait characteristics. This maintains a selection environment for those traits. But the prior existence of those preferences is not the ultimate explanation, and it is the *effects* of those preferences determining differential success in male breeding that provides the ultimate account. Put another way, female preferences, when *executed*, have an economic impact upon biotic interactions that in turn creates selection pressure. This is entirely in accord with Darwinian principles, and at no point did Mayr worry that intersexual selection was a special case of causation. Nor, more importantly, has he ever suggested that proximate *actions* cannot impact upon selection dynamics, contrary to Laland et al.'s later protestations (Laland et al. 2013).

The ultimate explanation for the overrepresentation of peacock tails of a certain kind is based on variation, heritability, and differential success, as usual. The sources of variation, heritability, and differential success are always proximate. More starkly, and following Mayr's discussion of Ho and Saunders (Mayr 1984), there is no need to understand how a female preference develops during ontogeny, nor to understand the underlying neurophysiology of those preferences in order to understand the competitive dynamics resulting from their execution, and the resultant selection. And Laland et al. make no effort to in fact do this (Laland et al. 2011). Displaying the proximate source of such dynamics would be interesting but entirely irrelevant to the evolutionary accounts and does not make a dent in Mayr's view about development.

There is another reason for skepticism about the reciprocal claim, as it has been outlined so far. It is perfectly possible to imagine a very large array of proximate architectures that might deliver the competitive dynamics at the core of the peacock

[8] But note, this is also because Laland and colleagues are thinking about form and in so doing siding with Pigliucci's call for a mechanistic account of form within evolutionary theory (Chap. 1, Pigliucci 2007).

[9] I am indebted to my father, David Dickins, for coining this term. It captures the age-old problem of which came first, the chicken or the egg? If this question were posed to an evolutionary theorist, they would say the egg, as chickens descended from dinosaurs. If it were presented to a developmentalist, they would also say the egg as they would be referring to a specific individual chicken. But if it were presented to an essentialist then there would be much dithering as she failed to find a reason to prioritize one or other essential type.

tail account. This is an issue of computation. Simply put, innumerable algorithms could be run on a peahen brain to deliver the preference bias observed in nature. It does not matter to the evolutionary story which of those computations is chosen. If the competitive dynamics are the same, then the trait bias will be the same in the population. Of course, when the authors established their example, they were not thinking at this level of detail and were in fact, thinking functionally.[10] They knew they needed a preference, and they knew at some point it would have to be proximately delivered. But they have not in fact cashed out any proximate details in their account so far. As said, their example conflates proximate and ultimate causation by allowing a form of functionalism into their descriptions of proximate activity, and this fails to find traction as a criticism of Mayr's distinction.

Laland et al. isolate what they consider to be a further limitation of Mayr's view stating that "modern causal modeling methods overcome Mayr's concern that biological complexity would make impossible an accurate description of causality as traditionally defined" (2011: 1512). In Sects. 4.2.3 and 4.2.4, I discussed how Mayr saw classical mechanics as a special and simplified case of causation that could be incorporated by modern statistical techniques. He was not concerned to render biology as classical mechanics but simply to point to the distinction in causal ambitions and to note the philosophical implications for the explanation. Laland et al. highlight directed graphs, structural equation modelling and the introduction of the information criterion to judge model fit as modern innovations and claim that this resolves Mayr's disquiet. That there was no disquiet makes this claim unusual, but it is weak because all of the techniques are in keeping with the statistical observations Mayr made. Where a directed graph might be used to map out alternative possible causal pathways, then their likelihood is statistically tested. Mayr's scrupulous use of the term statistical permits probabilistic and information criterion methods. What is more, as Laland et al. advertise, elements of path analysis are descendants of Sewall Wright's modelling innovations in evolutionary biology, innovations known personally to Mayr long before writing his 1961 paper.

4.3.2 Contemporary Debates

Laland et al. begin their reflections upon the role of causation in modern debates by characterizing those who align with Mayr's proximate–ultimate distinction as traditional, in contrast with the more radical non-Mayrian theorists.[11] Some radical thinkers see the proximate–ultimate distinction as a barrier to progress.

[10]For clarity, I am not using the term function here as Mayr does in his 1961 paper. Rather I am referring to a neutral brand of functionalism that is focused upon the role of mechanisms, rather than the mechanistic detail of their operation.

[11]This traditional-radical distinction strikes me as slightly premature in the great scheme of the history of science. To be radical is to suggest change, a break from the tradition. But to be a successful radical one has to make a truly alternative offer. The main thrust of my analysis is that

In support of the barrier argument, they quote West-Eberhard (2003), claiming that the "proximate-ultimate distinction has given rise to a new confusion, namely, a belief that proximate causes of phenotypic variation have nothing to do with the ultimate, evolutionary explanation" (2003: 11). But this quotation is taken out of context. West-Eberhard begins her section on the proximate–ultimate distinction clearly noting its intended use which is to avoid confusion between levels of explanation in biology. She goes on to state that:

> (I)t was an easy step from this important point to the idea that the mechanisms of development have nothing directly to do with evolution, or that they are the focus of a different research approach, one not primarily concerned with evolution and justifiably left aside by those primarily interested in selection and adaptation (West-Eberhard 2003: 10).

West-Eberhard agrees that there are different levels of explanation and that they are well captured by Mayr's version of the proximate and ultimate causation. However, she also believes that proximate mechanisms "are the causes of the variation upon which selection acts" (2003: 11). This point needs to be carefully positioned relative to those I made earlier (Sect. 4.3.1). In response to the intersexual selection case for reciprocal causation, I noted that the peahen preferences, when executed had an impact upon selection dynamics in order to clarify that it was the *effects* of those preferences that provided the ultimate causation. That effect was within differential success dynamics among males resulting from female choice rather than male competition. Differential success is one key component of natural selection, the others being variation and inheritance. Natural selection delivers ultimate causation. As I noted, all components of natural selection are proximately caused. West-Eberhard is also noting that variation is proximately caused in a book on developmental plasticity and evolution. This idea is not anathema to Mayr, who made clear the role of behavioral plasticity in evolutionary dynamics in his analysis of Lamarck (Mayr 1982, 2002). For Mayr behavioral plasticity allows solutions to contingent environmental problems, for example, and if those solutions positively impact upon fitness, then the underlying genes that enable behavioral response of the relevant kind will be favored. He ties this to the Baldwin Effect in his 2002 comments, something that West-Eberhard recruits for her own analyses later in her 2003 book.

4.3.2.1 Development and Evolution

It would appear that West-Eberhard does not see the proximate–ultimate distinction as a barrier to progress, but rather she sees the possibility for erroneous inference. But Laland et al. adopt a harder line and claim that the proximate–ultimate distinction was used to definitively separate development from evolution within the Modern Synthesis. They emphasize Mayr's discussion of DNA code, decoding of that code during development, and changes in the DNA code over evolutionary time, and they quote Mayr stating that all "of the directions, controls and constraints of the

criticism of the proximate–ultimate distinction does not amount to a true alternative and so this claim is, presently, somewhat overwrought.

developmental machinery are laid down in the blueprint of the DNA genotype as instructions or potentialities" (Mayr 1984: 1262).

As I have made clear (Sect. 4.3.1) Mayr's 1984 article was a review of a particular set of claims by Ho and Saunders, and Mayr did not see development as something to be excluded or separated from the Modern Synthesis. The instruction and potentialities quote is extracted from a section clarifying Mayr's commitment to development and evolution as separate processes, which is not the same thing as processes that cannot interact. Moreover, immediately preceding this quote, Mayr references epigenetic control of gene expression[12], which makes his subsequent comments on instruction and potentialities one about the context in which or over which such developmental processes work. In other words, Mayr clearly understands that DNA code has its effects modified by multiple processes on the way to expressing form during ontogeny. Put more pointedly, and if we are to follow the rhetorical line of Laland et al., then surely, we must ask whether or not they believe development can unfold in the absence of DNA code? If their answer is no, then a clear account of its role in each and every developmental situation is required, and that will amount to the kind of schematic, explanatory framework that Mayr has provided. That does not rule out different tokens of this kind of framework, but Laland et al. have not engaged at that level.

4.3.2.2 Niche Construction and Other Matters

Laland et al. discuss niche construction in terms of reciprocal causation.

> Niche construction theory, like developmental systems theory, is built on a reciprocal view of the interaction of proximate and ultimate factors. Niche-constructing (environment-altering) phenotypic traits of organisms coevolve with recipient traits via organism-modified factors in the environment. For instance, earthworms change the structure and chemistry of the soils in which they live and, by constructing their environment, modify selection acting back on themselves, for instance, influencing their water-balance organs. Here again there is reciprocal causation: The ultimate explanation of the earthworm soil-processing behavior is selection stemming from a soil environment, but a substantial cause of the soil environment is the niche-constructing activity of ancestral earthworms. Ultimate factors are not stable, autonomous features of the environment; they include labile features of the organisms themselves and their changing effects on their environment. (Laland et al. 2011: 1514)

Here again, we see reciprocal causation used not as an alternative to proximate and ultimate causation but as an account of the interaction between these logically separable processes (Dickins and Barton 2012; Scott-Phillips et al. 2011). Earthworm actions are caused by proximate mechanisms, and those actions have effects upon the environment, which in turn alter the selection environment as new

[12] Mayr's reference is to Waddington's 1942 conception of the epigenotype (Waddington 2012) which discusses the role of genes in influencing the many concatenated process of development. This is a pre-methylation, pre-histone modification, etc. view of epigenesis. It is clear from this paper that Waddington understands development to have important elements of genetic control, but also that it is not simply a situation of direct coding.

problems emerge. The new selection environment favors particular variants of extant proximate mechanism, and over time the population changes with regard to the representation of those particular phenotypic variants. Trait variation, differential survival and inheritance provide the standard ultimate level cause of evolution in earthworm populations; earthworm activity is a source of changes in differential survival dynamics.

Because evolution is a historical process, one can happily generate sequential descriptions of changes to variation, to differential survival and then map on changes in population characteristics. It is this kind of activity that leads Laland et al. to use the noun phrase *reciprocal causation*, where reciprocal is an adjective indicating that something is done in return (Stevenson 2010). Thus, proximate level actions lead to a change in differential survival through their *effects,* and this can lead to the selection of a different set of variants over time—ultimate causation at the population level is done in return for proximate activities. Pointedly, reciprocal causation fundamentally relies upon proximate and ultimate causation, as already argued, and is not itself a separate class of causation.[13] Mayr recommended the avoidance of noun phrases "that are nothing but reifications of processes" (Mayr 1982: 74–75), claiming that they were obfuscations that implied separate existence where there was none.

It is historical accounts of change that perhaps most interest Laland et al. Following their brief discussion of niche construction, they move onto human cooperation and cultural evolution, concluding with a remark about disciplinary boundaries:

> (C)ultural evolutionists view culture as a historical knowledge-gaining process and therefore as a legitimate source of ultimate explanations for acquired human characters. The ideas of cultural group selection or selfish memes illustrate in different ways the fact that researchers cannot safely treat culture merely as a proximate system, for the mechanisms link back to evolution through the nonrandom creation of new variation (Laland et al. 2011: 1515).

For Laland et al., historical transition is the same as evolutionary transition. Historical cultural change is underpinned by proximate learning mechanisms that allow discovery and social transmission. What is learned can be useful, can be transmitted, and can come to fixation within a population. These ideas could be said to conform to a form of General Darwinism (Chap. 1, Webb 2011) in that there is variation in ideas, inheritance of ideas and differential success of ideas. This is the basic conceit of meme theory (Dawkins 1976). If one is happy with General Darwinism, then one can draw a proximate–ultimate distinction within this field also. What Laland et al. reference in their paper is the idea that these General Darwinian processes can create new selection pressures that act upon the underlying genome, leading to biological evolution and the emergence of new behavioral traits in human populations. This

[13] It is possible that what really concerns Laland and colleagues is the idea that Mayr's view of natural selection, or ultimate causes, are in some way supervenient upon proximate actions and their effects, whereas they prefer to see them as emergent. This is never expressed in Laland's many writings on this topic, as far as I can see, but Mayr (1982) is avowedly in favor of emergentism.

might be so, but it is not a criticism of Mayr's model of causation, instead it again relies upon it in order to sensibly separate out levels of causation to understand the greater picture.

Just before the quotation above, Laland et al. state that evolutionary psychologists "characterize cultural influences on development as operating like a (proximate) switch (akin to the buttons on a jukebox) to shift behavior and cognition from one preestablished program to another, with each context-dependent strategy fashioned by natural selection" (2001: 1515). This is to be contrasted with the position of cultural evolutionists just outlined. But this is not contrast but rather a statement of the likely outcome of the kind of cultural processes Laland et al. endorse, which they claim to operate as a form of natural selection or ultimate causation. If social learning leads to the adoption of particular solutions to a problem, and that in turn creates a selection pressure for particular behaviors and their underlying proximate mechanisms, then once those behaviors are established within the population, they should respond to the appropriate cultural inputs in some conditional way. More bluntly, conditional architectures responding to inputs are not an entailment of Mayr's view alone.

4.3.3 A Hindrance to Progress

Laland et al. ask what role there is for Mayr's distinction in their concluding comments, and they suggest two components that can be differently viewed:

> The first component is that proximate and ultimate explanations should not be confused as alternatives. We agree. Biologists will always require different answers to how and why questions. The second component is that ultimate hypotheses cannot invoke proximate processes and are solely concerned with biological evolution. Here, we disagree. Progress within biology demands dismantling of Mayr's identification of proximate with ontogenetic processes and ultimate with evolutionary processes. (Laland et al. 2011: 1516)

Here, as elsewhere in the final paragraphs of the paper, the authors see some value in the proximate–ultimate distinction. But this value is tied very strongly to cases that they consider to have a unidirectional quality. Mayr's warbler migration example is such a case, where the external environment provides selection pressure for migration mechanisms, and once those mechanisms are in place, they do not change the environment in any significant way as a result of their actions. Such cases do not have the quality of reciprocation that Laland et al. see as more common.

The second component is a consequence of the focus upon reciprocal causation and the commitment to it being a true cause rather than an account of interacting causes. But the use of the term *invoke* is suitably vague. As I have argued, the causes of variation, differential success and inheritance are all proximate, as standardly understood under a traditional interpretation of Mayr and evolutionary theory. Any evolutionary account would at some point invoke the proximate processes involved in variation generation, etc. It is the case that ultimate hypotheses are solely concerned with evolution understood as a population-level effect. If Laland et al.

agree with this, then there is no real argument against Mayr's framework, and the use of reciprocal causation is at best a description of ongoing interaction across time that is typical of many situations. But if they do not agree with this, then they must be arguing that individual-level processes, such as development, which lead to change within the individual phenotype over a lifetime, are conceptually equivalent to changes in populations over historical time. Such a position is indeed a reversion to something approaching transformational thinking and a reversion based not on experimental data or any other empirical effort, but rather upon a commitment to particular metaphysics—that is, a belief that reciprocal causation is true causation and that it amounts to something more than the statistical interaction of proximate effects and the ultimate outcome of natural selection.

4.3.4 Philosophical Ambitions

My treatment of Laland et al. has thus far been fairly analytic, and I have sought to argue that their criticism does not present a clear departure from Mayr's causal framework and indeed heavily relies upon his distinctions. But there is another way to interpret their paper and to some extent, the comments of West-Eberhard that they cite. In their closing statement, Laland et al. emphasize an interest in progress within the biological sciences and claim that Mayr's work acts as a hindrance to this. They are suggesting that Mayr's work has the potential to cause a form of epistemic harm because it obstructs scientific success for individuals who are solely exposed to Mayr (Steup and Neta 2020). Thus, their critical review of the paper, half a century later, can be seen as an attempt to right that wrong. And to right that wrong means to enable biologists to realize that there are meaningful research traditions that have been developed within developmental biology, ecology (niche construction) and cultural evolution (this last explicitly related by the authors to the removal of another potential epistemic harm, through its commitment to multi-level selection).

A distinction can be drawn between scientific explanation and scientific understanding (de Regt 2017). For de Regt one can understand a phenomenon if one has a good explanation of it, but this further requires that one understands the relevant theory, which means that one is able to practically employ the theory. He sees understanding as akin to execution of a skill, the successful application of a theory to navigate a conceptual or empirical landscape (de Regt 2015). Understanding phenomena is what de Regt terms a macro-level aim for science, whereas understanding theory happens at lower layers and introduces the possibility of variation due to specific differences in the situations to which the theory is applied. These theories need to be intelligible if the macrolevel ambitions of science are to be achieved.

As an example, de Regt discusses behaviorism in psychology and the radical claim by behaviorists, such as Watson, that the aim of their science was prediction and control, something de Regt aligns with positivism because they sought lawlike relations to enable this. When mathematical laws proved impossible to derive behaviorists transitioned to positing unobservable constructs to explain stimulus-response relations, a position referred to as mediational neo-behaviorism (Moore

2013). De Regt claims that these mediational variables provided theoretical intelligibility that facilitated a functional explanatory framework that delivered prediction. The introduction of the micro-level theory provided the foundations for understanding. This leads de Regt to announce that only "intelligible theories allow scientists to construct models through which they can derive explanations of phenomena on the basis of the relevant theory" (2015: 92).

De Regt makes this last observation more formal with his Criterion for Understanding Phenomena, which states that a phenomenon is understood if and only if it has an adequate explanation based on an intelligible theory. Furthermore, that theory must "conform to the basic epistemic values of empirical adequacy and internal consistency" (2015: 92).[14] This package provides the necessary and sufficient conditions for scientific understanding.

The kind of causal complexity scientists are confronted by in the world is significant, with causal chains extended over infinity, multiple causes impacting upon a single event and causes interacting (Potochnik 2020).

> (F)aced with the need to grapple with this complex world, scientists face cognitive, and other, limitations. These limitations make it difficult to secure causal knowledge, to make accurate predictions, and to pursue science's other aims in this causally complex world of ours. Or, perhaps better, this point can be phrased positively: simple patterns are cognitively valuable. Simple patterns support human influence on and understanding of our world. There is thus a basic mismatch between the cognitive value of simple patterns and the world's complexity... So, in the face of this mismatch, we often resort to lying a little bit: we artificially simplify the parts of accounts that we are not interested in to improve our access in a variety of ways to the parts we are interested in. This is one service that idealizations provide. (Potochnik 2020: 934–935)

Potochnik explicitly aligns this use of idealization to render the world intelligible with de Regt's view of scientific understanding. For Potochnik, idealizations are false assumptions, or as she qualifies, assumptions made without regard to their veracity. They differ from idealized representations, which are to be seen as scientific models of the world that contain some idealizations, and something that de Regt notes can be far from accurate in truth terms (de Regt 2015). One particular use of idealizations is in the representation of causal patterns.

> Causal patterns are patterns insofar as they are regularities that are limited in scope and that may permit exceptions. The ideal gas law characterizes the approximate behavior of most gases, although its predicted relationships break down at low temperatures and at high pressures. It also ignores molecular size and intermolecular forces. Recall the idealization of an ideal gas composed of noninteracting point particles; this idealization achieves that neglect. Accordingly, even within its scope of application, the ideal gas law has exceptions... (T)o represent a causal pattern is to show how changes to a system would, over some range of circumstances, precipitate changes in other feature(s) of the system. The ideal gas law shows, for example, how temperature increasing in a sealed container of gas

[14] De Regt goes on to discuss criteria for judging intelligibility in terms of the ability of scientists to derive qualitative judgements about that theory without having to pursue exacting calculations.

> with a fixed volume increases the pressure. Mastery of causal patterns is exactly the kind of thing that beings who prize simplicity need in order to operate in and grapple with a causally complex world like ours. (Potochnik 2020: 935)

An intriguing element of Potochnik's analysis is that truth can be sacrificed in order to simplify, render intelligible and gain an understanding of causality. This activity differs from the accumulation of scientific knowledge, which is ultimately a factive[15] process grounded in truth (or attempts to attain veridical facts). This allows the possibility of scientific knowledge without scientific understanding, but Potochnik asserts that the reverse is not the case. While idealizations facilitate understanding through simplifying falsehoods, and the outcome is knowledge about causal processes understood in the delimited sense described above.

In light of this analysis, we can perhaps discern slightly different ambitions in the work of Mayr and Laland et al. In his 1961 paper, Mayr seems focused upon the intelligibility of biological theory, and he invokes two forms of causation to demarcate them from one another based upon his understanding of individual-level and population-level processes. Necessarily proximate and ultimate causation are presented as idealizations, something that is apparent when one compares his scheme to that of Tinbergen presented later in the same decade (Tinbergen 1963). Tinbergen separated the survival value of behavior from its phylogenetic history, regarding the former as a non-causal functional account and the latter as causal (Haig 2013). Evidently, this was more useful to Tinbergen in his quest for understanding within behavioral biology, which must have different aims to those of evolutionary biology as a whole. Or, in keeping with de Regt, the specific contexts of the use of evolutionary theory might allow for some variation in the idealizations that are deployed to gain understanding.

Laland et al. do not explicitly make their case within a disciplinary boundary, but their examples are drawn very specifically from intersexual selection, niche construction, developmental biology and cultural evolution. Their focus is upon the interaction of causes, and their idealization of this is captured by the concept of reciprocal causation. While I maintain that the very notion of reciprocity deployed requires both proximate and ultimate causation as separable and separate processes, following the analytic arguments above, we might charitably say that they are seeking understanding of particular kinds of interaction which become intelligible and more readily communicable[16] when packaged as reciprocal causation. Representing this as true causation of a third kind, instead of as an historical account of interaction, is a falsehood that might facilitate future science because it enables scientists focused upon interaction to deploy their skills better and to generate further knowledge in this domain. The strong claim of Laland et al., as at the beginning of this section, might then be that the traditional Mayrian approach can cause epistemic

[15] *Factive* denotes a verb that assigns the status of an established fact to its object (Stevenson 2010). In scientific accounts the verb *to know* is factive.

[16] De Regt (2017) makes the claim that his model of scientific understanding incorporates the core communicative aspect of science, as intelligibility implies communication.

harm if it prevents understanding of the interaction between the effects of proximate actions across time. Their reliance upon West-Eberhard (2003) to reinforce this point perhaps emphasizes that they really only see this as an available error, but their interpretation of Mayr's views on development suggests that they think he made it himself. I think this last point is not born out by Mayr's writings and amounts to a partial interpretation of him in light of Laland et al.'s ambition to generate specific understandings based on a technically false idealization of causation.

Potochnik's interpretation of this debate might still be simpler. She would see Mayr's causal framework as an idealization, capturing patterns and enabling understanding. She would see that as a form of knowledge. But she would also be aware that the idealization operates within limits. Her question to Laland et al. would be whether they had truly determined meaningful limits that led to a different understanding. The argument made here is that at best, their position flags an available error or omission with regard to historical interaction, but they have not proved that this error necessarily follows from Mayr's framework nor from his view of development, and therefore they have not added new understanding. So, their own is not a core challenge to evolutionary theory, and in particular the Modern Synthesis, but rather a cautionary note about interaction.

4.4 Subsequent Analyses

Haig has claimed that Mayr's 1961 distinction has led to two different idealizations (Haig 2013). He notes that Mayr established ultimate causes as a historical (phylogenetic) perspective that asked the question of a trait: *how come?* This translates as, how did this trait come to be; how did it arrive in this population? This was in contrast to the proximate questions around how it develops and operates within an individual. Haig believes that Mayr's critics have this in mind while Mayr's defenders align ultimate causation with the question: *what is the function of this trait*, or *what is it for?* This last is to be captured by inclusive fitness theory.

Haig suggests that the *how come* interpretation of ultimate causation is akin to Aristotle's efficient cause, the primary source of change, and he suggests that evolutionary change and development can both be seen as efficient causes. This is uncontroversial but neatly side steps Mayr's insistence on recognizing the distinction between individual and population-level thinking. In other words, because both can be classified as efficient does not mean that they operate in the same domains. Haig naturally suggests that the functional interpretation is akin to Aristotle's final cause, which aligns with some notion of purpose. As we have seen, Mayr takes an explicitly teleonomic view here, and natural selection is the cause of apparent design.

Tinbergen's desire to separate survival value (fitness) from his remaining three questions (mechanism, ontogeny, and phylogeny) due to function being seen as an effect of efficient cause is flagged by Haig, and he notes Tinbergen is not alone in this. I have argued a case for the effects of proximate processes already (Sect. 4.3.2), and so I find myself broadly agreeing with Haig's view that "'What for?' invites an 'on average' summary of past interactions between phenotypes and genotypes in complex environments. There is nothing in this view that denies an important role for development in evolutionary processes, nor that a recursive relationship exists among genotypes, phenotypes, and environments" (Haig 2013: 785). I think Haig is wrong to separate *how come* from *what for* because of this observation—he is right that natural selection, and thus teleonomy emerge from thinking about the evolutionary story of a trait. It captures the statistical interaction between variation, differential success and inheritance over time.

Haig's position does not equate development with evolution and instead sees an interaction between two processes as possible. And nor does it remove a role for deploying selection level arguments based on fitness maximization in order to derive explanations of traits. Indeed, the adoption of optimality modelling, based upon a functional causal interpretation, has proved extremely useful in generating explanations, predictions, and understanding (Potochnik 2008). Natural selection, as a functional account, may not please all as a causal story, but it is hard to deny its success as an idealization following de Regt and Potochnik (de Regt 2017; Potochnik 2020).

It is not clear that critics of Mayr entirely agree with Haig's view. Uller and Laland (Uller and Laland 2019) contrast Mayr's strong commitment to population thinking with *transformational* or developmental *explanations* in evolutionary biology (Sober 1984). Transformational explanations amount to aggregation accounts of a population, where individual differences are analyzed piecemeal to characterize a population structure. They differ from *variational* explanations, which rely upon capacities that get you into a population, in other words, selection-based accounts. As Sober points out:

> Lamarck's theory was developmental. It explained the evolution of species by laying down a sequence of stages through which life forms are constrained to pass. Species evolve because the organisms in them are gradually modified. In contrast, Darwin's theory of the evolution of species is not developmental. Darwin explained change in a species by a mechanism that permits (and.. even requires) stasis in organisms. In addition, the Darwinian paradigm views evolution as opportunistic, not preprogrammed. Selectional theories and developmental theories have fundamentally different explanatory structures. (Sober 1984: 148–149)

For Mayr, a population of migrating warblers is made up of individuals with the disposition to migrate due to selection. Evolution has occurred because there has been a change from an ancestral state of only a few birds with this disposition to the majority sharing it. The causes of individual variation in this disposition and how they developed are logically separable from those of evolution (Dickins and Barton 2012). Sober's dichotomy is perhaps best taken as an emphasis rather than insight. These are two parts of evolutionary biology, but not two parts of *evolution*.

Uller and Laland do not take this view and argue instead that the proximate–ultimate distinction has been used to carve a consensus about what constitutes satisfactory evolutionary explanations. They complain that the distinction makes "all of the sustained directionality in evolution (come) from fitness differences between genotypes, or natural selection" (Uller and Laland 2019: 3) and in so doing clearly see ultimate explanations in terms of *what for*. They emphasize the epistemic harm case by claiming that this outcome may inadvertently rule out legitimate alternative evolutionary explanations that may include developmental processes. These processes, they say, have been seen as irrelevant for explanations of phenotypic evolution (Uller and Laland 2019: 4). But phenotypes do not evolve; populations do.

Uller and Laland discuss killer whales (*Orcinus orca*) that socially learn how best to use local feeding ecologies (Foote et al. 2016). There is evidence that behavioral plasticity enabled whales to explore and adopt new ecological niches and then to develop foraging solutions through trial-and-error learning that were then passed on by social learning. Different groups of whales found different niches and became reproductively isolated. This isolation may have led to genetic drift and divergence in genotypes between the subpopulations. This is precisely the role for behavior that Mayr had in view when discussing Lamarck (Mayr 2002), but Uller and Laland locate such an account within *transformational* explanations and claim that a *variational* account would not invoke environmental influence on how traits originate and are inherited. This example makes clear the focus that these critics of Mayr have upon the emergence of *form* (Chap. 1), something in keeping with Pigliucci's call for a new mechanistic theory of form as a part of an extended synthesis (Pigliucci 2007). They do not ask, "is this population, so characterized, a product of the kind of change labelled *evolution*?" Instead, they wish to understand what causes individual trait variation.

The killer whale example conflates transformational and variational concerns or more precisely proximate and ultimate realms. This must be a consequence of adopting reciprocal causation as an idealization. A possible effect of this decision is that it seems to remove the need for an account of the selection for social learning. That account, under Modern Synthetic assumptions, would include a description of relevant ecological variability and the role of data led robustness solutions (Chap. 5). Uller and Laland address this criticism and suggest that such an approach would deliver an incomplete explanation, because it does not drill into specific historical contingencies for specific populations of killer whales. Moreover, it privileges genes[17] over development, and sees variation, differential fitness and heredity as autonomous processes. In so doing, Uller and Laland change the target of evolutionary theory and challenge its axiomatic structure.

[17] Throughout the various papers about Mayr, in the context of this topic, the fact that Mayr attacked bean-bag genetics as overly simplistic and reductionist, in part to protect biology from total reduction to chemistry, is never mentioned. The one thing Mayr's view of causation did not do is privilege the gene (Rao and Nanjundiah 2011; Smocovitis 1992).

4.5 Summary and Conclusion

Smocovitis (1992) positions Mayr's (1961) work on causation as a defense of the individuality of the discipline of biology. But this is a defense emerging from the unification of the field, via the synthesis, which saw the removal of metaphysical concerns and the foregrounding of empirical, axiomatic science. Given this, the critical attempt to replace proximate and ultimate causation with reciprocal causation is an attempt to step away from a core philosophical argument made toward the end of the Modern Synthesis period.

This interpretation is strengthened by the deliberate adoption of transformational thinking by Uller and Laland (2019) and their reliance upon an example that uses genetic drift, further distancing themselves from the hardened version of the Modern Synthesis that was firmly in place in the 1960s, when Mayr wrote his paper. They are not so much extending the synthesis, as rescinding it in favor of prior ideas. In effect, they are questioning whether some of the babies were thrown out with the bathwater during the synthesis. And in doing this, they are reverting to a focus upon the emergence of the form (Chap. 1).

While I have analytically tried to demonstrate that reciprocal causation relies upon proximate and ultimate causation to work, making claims for the replacement of those concepts false, I have also suggested that perhaps the idealization of reciprocal causation is of some use to particular research traditions. My intuition is that those will be disciplines in which agency is highly prized, given the clear introduction of agency as a cause of transformation (Buskell 2020; Walsh 2018).

References

Buskell A (2020) Synthesising arguments and the extended evolutionary synthesis. Stud Hist Philos Biol Biomed Sci. https://doi.org/10.1016/j.shpsc.2019.101244

Dawkins R (1976) The selfish gene. Oxford University Press

de Regt HW (2015) Scientific understanding: truth or dare? Synthese 192(12):3781–3797. https://doi.org/10.1007/s11229-014-0538-7

de Regt HW (2017) Understanding Scientific Understanding. Oxford University Press, Oxford

Dickins TE, Barton RA (2012) Reciprocal causation and the proximate–ultimate distinction. Biol Philos 28:747–756. https://doi.org/10.1007/s10539-012-9345-z

Dupre J (2017) The metaphysics of evolution. Interface Focus 7(5). https://doi.org/10.1098/rsfs.2016.0148

Foote AD, Vijay N, Ávila-Arcos MC, Baird RW, Durban JW, Fumagalli M, Gibbs RA, Hanson MB, Korneliussen TS, Martin MD, Robertson KM, Sousa VC, Vieira FG, Vinar T, Wade P, Worley KC, Excoffier L, Morin PA, Gilbert MTP, Wolf JBW (2016) Genome-culture coevolution promotes rapid divergence of killer whale ecotypes. Nat Commun 7(May). https://doi.org/10.1038/ncomms11693

Gould SJ (2002) The structure of evolutionary theory. The Belknap Press of Harvard University Press

Haig D (2013) Proximate and ultimate causes: How come? and what for? Biol Philos 28(5):781–786. https://doi.org/10.1007/s10539-013-9369-z

Lack D (1950) The breeding seasons of European birds. Ibis 92(2):288–316. http://onlinelibrary.wiley.com/doi/10.1111/j.1474-919X.1950.tb01753.x/abstract

Laland KN, Odling-Smee J, Hoppitt W, Uller T (2013) More on how and why: a response to commentaries. Biol Philos 28:793–810. https://doi.org/10.1007/s10539-013-9380-4

Laland KN, Sterelny K, Odling-Smee J, Hoppitt W, Uller T (2011) Cause and effect in biology revisited: is Mayr's proximate-ultimate dichotomy still useful? Science (New York, NY) 334 (6062):1512–1516. https://doi.org/10.1126/science.1210879

Mayr E (1961) Cause and effect in biology. Science 134:1501–1506. https://www.ifor.math.ethz.ch/education/bachelor/lectures/fs2011/math/Biomath/Mayr.pdf

Mayr E (1982) The growth of biological thought: diversity, evolution and inheritance. Belknap Press

Mayr E (1984) The triumph of evolutionary synthesis. Times Literary Supplement, November(2), 1261–1262

Mayr E (1992) Controversies in retrospect. In: Futuyma DJ, Antonovics J (eds) Oxford surveys in evolutionary biology, vol 8. Oxford University Press, Oxford, pp 1–35

Mayr E (2002) What evolution is. Phoenix

Monod J (1971) Chance and necessity. Vintage Books

Moore J (2013) Three views of behaviorism. Psychol Rec 63(3):681–692. https://doi.org/10.11133/j.tpr.2013.63.3.020

Pigliucci M (2007) Do we need an extended evolutionary synthesis? Evol Int J Org Evol 61 (12):2743–2749. https://doi.org/10.1111/j.1558-5646.2007.00246.x

Potochnik A (2008) Optimality modeling in a suboptimal world. Biol Philos 24(2):183–197. https://doi.org/10.1007/s10539-008-9143-9

Potochnik A (2020) Idealization and many aims. Philos Sci 87(5):933–943. https://doi.org/10.1086/710622

Rao V, Nanjundiah V (2011) J. B. S. Haldane, Ernst Mayr and the Beanbag Genetics Dispute. J Hist Biol 44(2):233–281. https://doi.org/10.1007/s10739-010-9229-5

Scott-Phillips T, Dickins TE, West SA (2011) Evolutionary theory and the ultimate–proximate distinction in the human behavioral sciences. Perspect Psychol Sci 6(1):38–47. https://doi.org/10.1177/1745691610393528

Scriven M (1959) Explanation and prediction in evolutionary theory. Science 130(3374):477–482

Smocovitis VB (1992) Unifying biology: the evolutionary synthesis and evolutionary biology. J Hist Biol 25(1):1–65. https://doi.org/10.1007/BF01947504

Smocovitis VB (1996) Unifying biology: the evolutionary synthesis and evolutionary biology. Princeton University Press

Sober E (1984) The nature of selection: evolutionary theory in philosophical focus. University of Chicago Press

Steup M, Neta R (2020) Epistemology. In The Stanford Encyclopedia of Philosophy (Fall). Metaphysics Research Lab, Stanford University. https://plato.stanford.edu/archives/fall2020/entries/epistemology/

Stevenson A (ed) (2010) Oxford dictionary of english, 3rd edn. Oxford University Press, Oxford

Tinbergen N (1963) On aims and methods of ethology. Zeitschrift Fur Tierpsychologie 20:410–433

Uller T, Laland KN (2019) Evolutionary causation. In: Uller T, Laland KN (eds) Evolutionary causation: biological and philosophical reflections. MIT Press, London, pp 1–12

Waddington CH (2012) The epigenotype. Int J Epidemiol 41(1):10–13. https://doi.org/10.1093/ije/dyr184

Walsh DM (2018) Organisms, agency and evolution. Cambridge University Press

Webb RH (2011) If evolution is the answer, what is the question? J Evol Psychol 9(2):91–107. https://doi.org/10.1556/JEP.9.2011.2.2

West-Eberhard MJ (2003) Developmental plasticity and evolution. Oxford University Press, Oxford

Data and Information

5

Abstract

The concept of information has been widely deployed within biology. In this chapter I argue that this has largely been done in a colloquial manner, leading to a lack of clarity. In this chapter I argue for a definition of information as the functional relationship between input and system and against the idea that information is something to be captured and used by a system. This position enables greater clarity about key aspects of biology that have been the focus of criticisms from advocates of the Extended Synthesis, and my strong claim is that this view of information is implicit within Darwinian and Modern Synthetic thinking. To make this case, I apply this version of information to an account of plasticity, which marks the first formal discussion of developmental biology and lays the groundwork for subsequent chapters dealing with the developmental challenge to the Modern Synthesis.

Keywords

Information · Data · Context · System · State change · Surprisal · Autopoiesis · Preformation · Gene centrism · Content · Representation · Levels of organization · Mechanism · Robustness · Bet hedging · Developmental plasticity · Physiological Plasticity

5.1 Introduction

In this chapter I introduce a key distinction between data and information that will inform subsequent commentary and discussion in the remaining chapters. My core claim is that this distinction will bring greater clarity to many programmatic statements from the Modern Synthesis. Related to this, I argue in the closing section that this view of data and information is fundamentally a view of the central problem of biology, and while Darwinian thinkers have never explicitly recognized this, to

T. E. Dickins, *The Modern Synthesis*, Evolutionary Biology – New Perspectives on Its Development 4, https://doi.org/10.1007/978-3-030-86422-4_5

the best of my knowledge, the development of the evolutionary theory was the realization of this idea. In passing, I also apply this distinction to the concept of robustness, and its complement, plasticity. In doing this, I begin to develop a position on the relationship between evolution and development, which is fully expressed in Chaps. 6 and 7.

5.2 Data and Information

Information is a property of a system where the system can be in more than one state, and the nature of the state it is in is determined by some kind of input. Thus, if a system can exist in S_1 to S_n states, that input that causes it to change from S_1 to S_2 is *information*. Or more precisely, it is *informative*. This is the standard derivation from Shannon information that is common currency in biology (Godfrey-Smith 2007).

The etymology of information "connotes a distinctive arrangement" (Cohen 2000: 13) or form, an idea that arose from Shannon's attempts to improve fidelity in telephone transmissions, such that the output matched the input to a high level (Shannon 1948). To do this, Shannon required a measure of the degree to which the message conformed to a specific arrangement or form, something Cohen refers to as being *just-so*. Shannon hypothesized that the degree to which an arrangement is *just-so* is related to the other possible arrangements of the relevant elements of the message, which can be understood as possible errors. So, if the message in question is a codon—CAG—the other possible permutations are CGA, AGC, ACG, GCA, and GAC. As Cohen notes, the *just-so-ness* "stands out against a background of the alternative arrangements, the possible 'errors' that may have been constructed out of the sub-units of the message. Just-so-ness thus has a quantitative aspect; the more possible 'errors' there might be, the greater the just-so fidelity of the message" (Cohen 2000: 14).

Cohen further notes that the more possible errors there are then, the lower the probability of the *just-so* arrangement arising, and when it does arise, it is surprising. More technically, this property of information is referred to as *surprisal* (Clark 2013, 2018). Surprisal measures the probability of an input occurring given some particular context. A high surprisal input is an input with a low probability of occurrence, and therefore that input has a greater informational role (is more informative) than low surprisal inputs, which are highly predictable given the context.

In relation to communication, certain inputs can be more or less likely. Cohen contrasts a telephone connected directly to your mother's telephone with one that is connected to all telephones in the world. In the first case, when the telephone rings, and it is your mother at the end of the line, there is no surprisal and no information, as the probability of it being your mother was 1.0. In the second case, it could be anyone calling you when the telephone rings, and so when the caller identifies him or herself, the surprisal value is high and extremely informative. Cohen notes that patterns of caller history will change the probability distributions, and so regular calls from your mother on the second telephone will carry less information or surprisal value than the caller identifying herself as Angela Merkel.[1]

How does this communicative view of information relate to my opening comments about systems and state change? When the telephone rings in the second scenario, then the uncertainty of the person reaching to pick up the telephone is high. Once the caller identifies, the recipient of the call has her uncertainty reduced. Various other aspects of her behavior can thus unfold now she knows she is talking to Angela Merkel. So, we can say that if a system can be in *n* states, then the precise or *just-so* input that leads to state change reduces the uncertainty of the system and that input becomes information with a surprisal value. While this is a property of the probability distributions of possible inputs, it is also a property of the system itself that is prepared to take those inputs. Put another way, how the system can change states determines, in the sense of defines, the *just-so* nature of the input. The precise or *just-so* nature of the input is a property of the *relevant* alternative configurations of the system.

Floridi helps us to fill this position out in a little more detail. He considers Shannon's theory to be about data rather than information *per se* (Floridi 2003, 2010). This is because he adopts a General Definition of Information that states that *Information = data + meaning*. By meaning, Floridi "means that the data must comply with the meanings (*semantics*) of the chosen system, code, or language in question" (Floridi 2010: 21). Floridi is making clear that Shannon's mathematical theory of communication gave a method of measuring surprisal as a function of the probability of a particular input, but it made no statement about the nature of the systems into which inputs arrive. Those inputs are data. However, the system, as just described, is configured in a way to accept certain inputs, or data, albeit on a probability distribution, and to change states upon receipt of those inputs (technically reducing uncertainty). The system configurations provide context, or meaning, for the data. Given this I prefer the formulation: *Information = data + context*.

Boisot and Canals see traditional and colloquial uses of information as essentialist[2] and "of a concern with *being* rather than with *becoming*" (Boisot and Canals 2004: 46). Here they are referring to those treatments of information that see it as something out-there to be harvested, or taken, and then used. They are clear, with Floridi, that information is not a thing, but rather it is a relation. One reason they claim this is because of the relationship between data, cryptography, and information. They tell us that a database can be made publicly available but encrypted such that only those with the key to unlock that encryption can use the data to inform them. "Cryptography, in effect, exploits the deep differences between data and information" (Boisot and Canals 2004: 43).

---

[1] To the best of my knowledge, Angela Merkel, the Chancellor of Germany at the time of writing, has no children.

[2] This is in keeping with Popper's view on essentialist definitions (Popper 1945) such that colloquial usage assumes that there is information and seeks to understand it, rather than applying the term *information* as a short hand for a list of defined phenomena (the nominalist approach to definition).

Boisot and Canals tell us that *data* "can be treated as originating in discernible differences in physical states-of-the-world—that is, states describable in terms of space, time, and energy" (Boisot and Canals 2004: 46). Their view is that anything that can be discerned is grounded in such differences and that it is agents that discern. They note that agents are immersed in stimuli from the physical world, but to discern stimuli as *data* relies upon the actions of neurological mechanisms that will register stimuli when an energy threshold is exceeded. They explicitly see this as a filtering system, and once through the filters, stimuli are registered as *data*. What Boisot and Canals refer to as "significant regularities" (Boisot and Canals 2004: 47) in the *data* can be extracted by agents as *information*. What counts as a significant regularity is a function of the agent's disposition, according to this view. "To summarize, we might say that *information is an extraction from data that, by modifying the relevant probability distributions, has a capacity to perform useful work on an agent's knowledge base*" (Boisot and Canals 2004: 47). Knowledge amounts to a set of expectations based on prior learning exposures that can be updated by *information*.

Boisot and Canals come close to the view I am advocating in this book. My view is that *data* are stimuli but that there are two kinds—usable and unusable. Both are *potentially* usable physical stimuli emanating from the world, and that potential can be realized during the process of evolution. So, what makes some data usable is the existence of systems that can take those stimuli as inputs. Taking stimuli as inputs amounts to something like the filter view from Boisot and Canals and is minimally mechanistic, such that a semi-permeable cell membrane permitting certain ion transfers and not others can be seen as a filter. Once stimuli can enter a system, they can affect state changes. Passing the filter makes stimuli into *usable data*; subsequent responses to that data are a property of the contexts provided by the system. It should be noted that I am using the system to refer to the organism (or agent) as a whole while also assuming the organism is in fact a collection of sub-systems, each with their own informational character due to specific *data + context* relations.

It is not quite clear what Boisot and Canals mean by discernible differences in the physical states of the world. If the property of *discerned difference* is being attributed to *data*, it assumes a certain amount of processing, beyond mere filtration, that would rely on systemic state change (and thus on the neurological mechanism). Indeed, it suggests data carries content and lends data an informational flavor, as the input within a context is causing an effect. *But* Boisot and Canals may wish that we interpret *data* as making discernment possible, but that this activity is a property of the systems into which the data enters. This last is clearly my preferred interpretation of their work, and I think in the spirit of their paper. Thus, the view of information from Dretske, that relies on the correlation between A and B such that B carries information about A under a conditional probability structure, can be maintained if we interpret that correlation as a process derived within a subsystem of the agent

(Dretske 1983). It is not a property of the signal but a property of the relationship between *data* and systemic *context.*[3]

A biologist taking the external view when trying to decipher the data-to-context relation will be operating at a particular level of biological organization and regarding that as a system. But that system is also a context in which data has an effect. A principal method of beginning this task is to look at the function of the system—what work does it achieve? So, the biological use of Shannon is to claim that biological systems can take a finite array of inputs and change state accordingly and that the entirety of that package is informational, or rather, *the interaction of relevant inputs with relevant biological systems creates information.* For biologists, the focus is upon identifying *relevance* in this statement.

5.2.1 Discerning Data and Information

Shannon information amounts to no more than a description of what information is, or more precisely, what data is and does (Boisot and Canals 2004; Floridi 2010). With this view of data in mind, we can then set about trying to understand the contextual structures of systems that make use of specific data. In one sense, then, we can argue that the job of biologists is to discern biological meaning, where meaning is synonymous with context. In doing this biologists, and other scientists, are making an ontological commitment to a relationship, to the existence of systems that take input.

This ontological commitment is intuitively grounded in what we might see when doing natural history. From the most basic interaction between a predator and its prey, through to clear use of signals for communication. Shannon's quantitative insights were, after all, derived within a communication project. But the commitment also relates to an autopoietic concept of life.

> An autopoietic unit is a system that is capable of self-sustaining owing to an inner network of reactions that re-generate all the system's components... In other words, an autopoietic system organizes the production of its own components, so that these components are continuously re-generated and the system can therefore maintain the very network process that produces them... In this way, autopoiesis is capable of capturing the mechanism that generates the essence of the living. The components organize themselves (auto-organization) in a bounded system that produces the components that in turn produce the system, and so on. (Luisi 2003: 51)

Autopoiesis was developed as an abstraction from the cell because cellular life was the only form of life known on earth, a kind of natural historical observation

[3] Another reason to firmly place correlations as an outcome of systemic properties interacting with data is that as all data are physical stimuli there is a huge collinearity problem making real correlation useless. Filtering data reduces this, affecting what is later attended to. But the nature of the how inputs are related and change states in systems is impacted by selection, such that those systems that maximize fitness by deriving useful correlations thrive.

(Ginsburg and Jablonka 2019). In this way, autopoiesis is an abstraction from a particular observation set and all the commitments that might entail. This cellular view gave autopoietic units boundaries within which various processes occur that allow regeneration and homeostatic maintenance. While the cell membrane is semi-permeable, allowing specific chemicals and nutrients to pass within from the external environment, the unit is effectively self-sustaining. The input of energy through the semi-permeable membrane to replenish diminishing resources that enable self-sustenance makes these systems physically open. But, in spite of this ontological openness, autopoietic systems are logically self-contained and have what has been referred to as an epistemic closure (Luisi 2003). Clearly embedded within autopoiesis is a notion of autonomy delivered by the internal complexity of the unit (Gershenson 2015).

It would be possible to treat autopoiesis as a direct, if minimal, definition of life and to treat it as some kind of essential natural kind (much to Mayr's distaste). But as with certain interpretations of closely related ideas (Griesemer 2015) I think it is better to see autopoiesis as a useful model or heuristic to borrow Griesemer's term, which helps to guide scientific development. For example, it is possible that we might grade the degree of autonomy that systems have, perhaps as a function of internal complexity, demonstrating that some are more autopoietic than others (Gershenson 2015), even though autopoiesis was originally established as an all or nothing concept (Luisi 2003). This might make particular sense when pursuing phylogenetic questions about the origin of life as well as core concepts about the emergence of individuality (Maynard-Smith and Szathmary 1995; Queller and Strassmann 2009). For these kinds of enquires, we might expect blurred boundaries of historical transition as solutions emerged.

The autopoietic heuristic enables us to think about multicellular organisms also. Each of those cells in a multicellular organism takes energy inputs to facilitate internal processes of maintenance and regeneration. But the overall organism also takes other inputs to actively maintain its autonomy and can be regarded as an autopoietic unit. These inputs will include sensory inputs that enable the detection of external threats and energy resources, as well as navigation to avoid threats and find resources in non-sessile organisms. The increase of permitted inputs in some way matches the external complexities with which the autopoietic system must interact in order to remain autonomous and not come under total environmental control. The greater the external complexities then, the more complex the internal context for any inputs, and the greater the informational capacity of the organism (cf. Godfrey-Smith 1996, 2017).

We see these basic assumptions at play in scientific practice. If we think about an experimental intervention, at the proximate level, we might characterize that statistically, in keeping with the general linear model, such that $Y = f(x)$. What this means is that we look to the effects upon a dependent, or response variable, Y, in relation to an independent, or input variable, x. If we manipulate x to some extent, holding everything else constant, how much variation in Y does that cause? The f refers to a mathematical function so that after experimentation or measurement (under a correlational design), we know the intercept and slope of a fitted line, as well as the residual error on either side of that line. Thus, Y varies in relation to x under those functions, and x predicts Y to some level of probability as a result of f. The general linear model, in this example, nicely captures Mayr's comments on the statistical nature of causation in biology (Sect. 4.2.3) (Mayr 1961).

The relationship between the independent and dependent variables in the experimental context is a system-level relationship between an input, x, and a response, Y, which amounts to a state change in the system. While these experiments are often conducted under very specific hypotheses, designed to titrate effect sizes and interactions with other variables and so forth, the baseline theoretical assumption is that the input is being tested for relevance to the system. If x does indeed shift Y variance, then the assumption is that x has played an informational role in the system. If it does not, then the assumption is that x has no role in the maintenance or regeneration of Y. All of these initial assumptions, of course, warrant further exploration and testing.

It should be noted here that the original notion of autopoiesis itself does not rely upon the view of data and information that I have presented, not least because the original notion explicitly did not require these concepts in order to characterize life (Cárdenas-García 2020). But I think that data and information in the context of biology are logical entailments of the autopoietic view of life from the point of view of science. More firmly, autopoietic life creates information, understood as a functional relationship between an input and a system. What this means is that biologists can use information talk to frame and organize their enquiries about how and why organisms are the way they are without committing to an essentialist natural kind view of information. Bluntly put, it is the fact of organic life that creates these relationships: life is context.

5.2.2 Information, Preformation, and Genes

There are three common and related dichotomies that have been proposed within biology: nature/nurture, innate/learned, and robust/plastic (Bateson and Gluckman 2011). It is my view that were we to treat these as true oppositional dichotomies, we would be in error. Thus, to think of nature *versus* nurture when trying to account for a particular trait would be a mistake because nurture is something that is found in nature and therefore nurturing behaviors are part of the set of naturally occurring things. Similarly, learning is the function and product of learning mechanisms, which are not themselves the product of learning. The properties of such mechanisms will permit certain kinds of things to be learned. Plasticity, as I will discuss in more detail below, is one way of being robust, for it enables an organism to weather the vicissitudes of life.

Nature *versus* nurture opposes something fixed, innate and fully formed at birth with something flexible and acquired and in need of learning and development. This is often parcelled into the oppositional dichotomy of *nativism* versus *empiricism* along with innate/learned and robust/plastic. This dichotomy has a long history dating back at least to Pythagoras and the distinction between *preformation* and *epigenesis* (Coall et al. 2014, Chap. 7). There were various preformation theories during this period, including the homuncular theory that mooted a small but fully formed adult present within the paternal or maternal substrate. Preformationism, then, is the nativist idea that all traits have a complete form at conception when male and female contributions are combined. It was Pythagoras' view that the male provided the essential contribution and the female an environment for growth. Aristotle promoted epigenesis in which *material* causes were contributed by both parents, thereafter *efficient* causes led to the emergence of the *formal* cause, which has *final* purpose. This is an empiricist idea where form emerges over developmental time.

Preformation implies that the distinct form of an organism is already given. All that is needed is growth to attain the adult size and presumably reveal full functionality. There is no role for input here and no role for context. As such this view is not consilient with the definition of information, I have so far pursued. Indeed, this view is perhaps more similar to the colloquial uses of information. For example, when you tell your traveling companion that you are going to the hotel reception to get *information about* local restaurants, there is a working assumption that information is something in the world to be harvested (Boisot and Canals 2004). In reality, the interaction you have with the receptionist conforms to the *data* + *context* model, as the receptionist reduces your uncertainty about places to eat. But in everyday terms, we *capture* information, and this is because we tend to focus upon a representational end point, which minimally is a change in our internal contexts with no need to commit to a theory of those contexts.[4] And it is also because we tend to refer to information in intentional language, where information is *about* or *for* something. In doing this, we package the entire *data* + *context* effect and conflate it with the mere input that we seek.

Preformation theories probably have their origins in this way of thinking, in the absence of a modern scientific biology. If all we know is that sexual behavior is closely associated with reproduction, and reproduction tends to produce organisms that are similar to the parents, then we can see sexual reproduction as a behavior that somehow marshals information about human bodies. In this particular case, that information is also seen as the distinct arrangement associated with parental (and species) form. It is not hard to see that this view is an essentialist one, and following from this, it is not hard to see that our everyday *information-as-out-there* talk is also essentialist. There are lots of problems with essentialism, but a crucial one to note here is that preformation, interpreted in this way, does not provide a satisfactory causal argument for the development of form. Similarly, a view of essential information out in the world to harvest does not explain how we use that information.

Epigenesis might appear to be a better option given the preceding argument. Some initial conditions are provided that lead to development within a particular causal context, following Aristotle. But it is possible to adopt this view under what

[4]Maximally, for Boisot and Canals (2004) it would be an updating of knowledge understood as prior learning history.

we might call a natural theological perspective and simply claim that those processes were all put in place by God and follow a divine plan. It is this plan that can deliver essentialism because it is a representational instruction (Oyama 2000).

A persistent criticism of the gene-level view of evolution that emerged during the Modern Synthesis was that it was gene-centric (e.g., Noble 2011; Oyama 2000; Sultan 2019). In part, this is a criticism based on privileging the gene in causal accounts of both development and evolution (Dickins 2020). Oyama, in particular, has related the replicator concept of the gene to preformationism, arguing that some accounts see genes as containing the specific instructions for development and that those instructions have been placed within the gene by natural selection. She claims that biologists at times "*assign formative relevance only to the DNA, where the encoded representation of the phenotype (or of the instructions for building it) is thought to reside*" (Oyama 2000: 30). She discusses the idea that natural selection has simply replaced God in the natural theological mind, and this new force has given genes "ultimate formative power" (2000: 31).

For Oyama, a part of the problem seems to be how generalities or abstractions are constructed. As an example, she discusses Monod.

> Jacques Monod, who, with Jacob, presided over some of the most exciting developments in early molecular biology, describes in minute detail various macromolecular processes and their organizing functions. When he engages in straightforward description, the complexity and interdependence of causes are clear. When he interprets these processes in more general terms, however, an interesting thing occurs. He says, for example, that the genome "entirely defines" protein function, and asks if this is contradicted by the statement that the protein's three-dimensional structure has "data content" that is "*richer* than the direct contribution made to the structure of the genome." His answer is that, because the three-dimensional, globular structure appears only under "strictly defined initial conditions," only one of all possible structures is realized. "Initial conditions hence enter among the items of information finally enclosed within the globular structure. Without specifying it, they contribute to the realization of a unique shape by eliminating all alternative structures, in this way proposing—or, rather, imposing—an unequivocal interpretation of a potentially equivocal message." (Oyama 2000: 31–32)

Oyama's complaint with this account is that while Monod admits the initial conditions are among many items of information in the final, folded protein, he also claims that those initial conditions specify the final, folded form. Oyama marks this as an inconsistency or contradiction within Monod's account. I think she is right that Monod has painted himself into a corner here, but I think that is the case because of the colloquial use of information, as discussed above, and also because of his treatment of *data* as synonymous with *information*. When we use *information = data + context* we can repackage Monod's account. DNA provides data to a protein synthesis context, and it is reasonable to see those data (codons) as initial conditions that then interact with RNA and ribosomes to produce polypeptide chains that are folded. RNA and ribosomes provide context, but RNA also becomes an input. There is more to the process than this, but the point is that inputs into contexts can produce subsequent inputs into other contexts. Many theorists tend to think in hierarchies of organization at this point (but see Sect. 5.2.3).

To reinterpret Monod—a folded protein is the outcome of multiple data inputs into an overall protein synthesis context. That overall context is actually compartmentalized and organized in order to allow data flow. Initial starting conditions will trigger conditional responses in sub-contexts via data input—and the contexts are prepared to take an array of inputs and to respond systemically. But the initial data inputs will not fully determine the ensuing process, and they will just make particular outcomes more likely. Thus, there is internal regulation that is sensitive to contextual factors. An example of this is attenuation in bacteria, where transcription can be halted in response to available amino acids on transfer RNA, which allows time for the synthesis of the required amino acids (Henkin and Yanofsky 2002). This is an entirely conditional architecture, dependent upon data inputs, that operates in a context further downstream from the initial starting condition inputs. As the process runs, the probability of a particular protein form emerging increases, but at each point, along the way, data will have surprisal values with relation to that final form.

I think Monod's claim that the genome entirely defines protein function is a statement about the necessary role of genetic input. It would be wrong to claim it is sufficient, and Oyama concedes the detail of his account makes plain he does not think this. Using this to motivate a principle of charity, we might reinterpret the statement as a counterfactual argument. Under this interpretation, Monod is stating that protein synthesis would not be possible without genetic input and that the existence of the other elements of the overall system depends upon that input both in ontogenetic and phylogenetic terms. More baldly, Monod's rhetorical claim forces us to ask whether or not we can envisage protein synthesis without genes (understood in replicator terms). Most biologists would answer "no," and at this point would give genes conceptual but not ontological primacy in any detailed account. Moreover, it is clearly not a claim for the representational and instructional completeness of the gene—indeed, the gene has no instructional role under this view of information. Instead, this interpretation amounts to an organizing principle for biology, and one that does not clash with let alone refute Oyama's deep and useful interest in developmental systems and interdependence.[5]

My point is that the concept of information needs to be used carefully in order to develop organizing principles for explanations and that caution is derived from a particular commitment to what information is. But I also noted that many biologists tend to think in hierarchical terms with regard to organization, and this may introduce problems also.

[5]It is easy to align Dawkins' catalytic view of the gene with this kind of account also (Dawkins 1989), where genes are initial inputs but their impact on the distal protein structure is necessarily indirect.

5.2.3 Levels of Organization

It is common to assume that biological organization is hierarchical. For example, within a body, we can think of genes that are parts of cells, cells that are parts of organs, organs that are parts of individuals, and individuals form groups. This particular example of a hierarchy is one of the entities. Recently, Eronen has listed the reasons for recruiting levels of organization (Eronen 2015). Here are three that are specifically pertinent to the discipline of biology:

1. To enable reductive explanation.
2. To capture and understand aspects of nature.
3. To analyze top-down causation.

All three have an explanatory role, but option 2 also includes an ontological ambition, and the opening entity-based example is a case of this kind. Eronen discusses a proposal from Craver and Bechtel that applies levels to biological mechanisms in a way that will permit ontology, explanation and also an account of top-down causation (Craver and Bechtel 2007).

Mechanisms can be thought of as generating phenomena as a consequence of the coordinated activity of their component parts, such that the actions of one component will enable the actions of another (Bechtel and Bich 2021). Bechtel and Bich offer a refinement to this view stating that "mechanisms are systems of constraints that restrict the flow of free energy to perform work" (Bechtel and Bich 2021: 2), with constraints working against dynamical relations. Biological systems rely on constraints to direct energy and fight entropy, those constraints being boundaries that contain dynamical processes. Moreover, according to Bechtel and Bich, those constraints are unaffected by the processes they enable:

> An enzyme is an example in which constraints lower the activation energy necessary for a reaction, thereby catalysing the production of an otherwise improbable product, while not being consumed in the reaction. At a different scale, the vascular system operates similarly to pipes. It canalizes the distribution of blood towards specific organs, which could not be accomplished by diffusion alone. As a result of constraints, production mechanisms function to synthesize, repair and replace components of the organism and generate its basic activities and behaviours. What is distinctive of biological organisms is that these constraints and the mechanisms they constitute collectively contribute to maintaining the conditions for continued existence, thus realizing a causal regime called 'closure of constraints.' In this view, living systems are autonomous; they maintain themselves and self-maintenance is their ultimate norm. (Bechtel and Bich 2021: 3)

Craver and Bechtel claim that whole mechanisms exist at higher levels than their component parts; thus scientists decompose mechanisms that deliver a phenomenon into their constituent parts (Craver and Bechtel 2007). Levels of organization apply within specific mechanisms, and one can find sub-mechanisms all the way down as they illustrate with the following example:

> The eye is a familiar and unproblematic example of a multilevel mechanism. At the highest level, the eye transduces light into a pattern of neural activities in the optic nerve. This process can be decomposed into lower-level components and their activities. The light enters the eye, it is inverted and focused by a lens, and it is projected onto the retina, where the information in the light is converted into a pattern of neural activity in the optic nerve. The conversion of light into patterns of neural activity by the retina can itself be decomposed into different components: in particular, the rods and cones that change their electrical state depending on specific features of the light stimulus (such as wave- length and intensity). Another level down, rod cell activation is also sustained by a mechanism. Light is absorbed by and activates rhodopsin, which then stimulates G-proteins. These G-proteins activate cyclic GMP phosphodiesterase, which catalyzes the conversion of cyclic GMP to 5¢-GMP. Lowering the concentration of cyclic GMP causes sodium channels to close, reducing the inward sodium current and thereby hyperpolarizing the cell... Each new decomposition of a mechanism into its component parts reveals another lower-level mechanism until the mechanism bottoms out in items for which mechanistic decomposition is no longer possible. (Craver and Bechtel 2007: 549)

Craver and Bechtel are keen to note that their own is a local approach to specific problems and that it is a constitutive and mechanistic approach designed to make sense in terms of the business of biology. Here they are explicitly removing their account from those philosophical positions that allow emergent properties. Emergent properties are, by definition, not explicable in terms of the mechanistic causation at the level below, which Craver and Bechtel treat as a concept somewhat alien to the ontological commitments of biology.

The characterization of multiple levels of mechanisms in the preceding example demonstrates the complexity, but it does not commit Craver and Bechtel to a global organization structure across all mechanisms. This hierarchy works in this instance but implies nothing about the outcomes for studies of other phenomena. What is meant here is that any commonality with another set of mechanisms is an open empirical matter but not an ontological conclusion that can be derived from understanding this particular, local set-up. In this way, Craver and Bechtel have a minimal view of levels compared to the more grandiose schemes that saw disciplinary boundaries as effectively marking large-scale horizontal boundaries in the ontology of the natural world (Oppenheim and Putnam 1958).

Eronen (2015) claims that Craver and Bechtel's talk of levels is misguided. He notes that a particular component of one mechanism (he uses NDMA receptors in synapses involved in delivering long-term potentiation) can be found in other mechanisms, but that there is no coherent sense in which one might talk of an NDMA level in biology. Instead, Eronen prefers the idea of a branching hierarchy of relationships, not least because something like NDMA can be found operating in neurons at very different levels in the componential structures of other mechanisms.

That NDMA or any other component can be found at work in multiple mechanisms makes componential relations insufficient for a full explanation of a given phenomenon. Eronen suggests introducing a concept of scale in order to note the relations between things, and he mentions size and time as two useful starting points. What is more, he points out that the kind of levels of organization typically referenced in biology, such as the opening entity-based example, incorporate scale (in this case, size) quite naturally. His claim is that levels-talk does not assist in delivering aims 1 and 3 above. Reduction is more readily delivered simply by focusing upon componential relations as well as upon scale. For example, psychological processes happen at a slower time scale than the neurophysiological processes delivered by their component parts. This leaves the ontological status of levels in doubt, and Eronen sees them as merely heuristically useful when approaching a particular biological problem, because they capture details of componential organization and scale for that particular issue.

Eronen's analysis preserves Craver and Bechtel's focus upon mechanisms in order to deliver explanation and a brand of reduction. Mechanisms have been defined in relation to the coordinated activities of their components, the restriction of free energy, and the production of phenomena. Moreover, this view of mechanism is a principal component of autonomy, which is linked to minimal life characterizations and the concept of robustness (Bich 2018) (Sect. 5.3). However, Eronen denies levels an ontological status leaving only bespoke epistemic application for such talk.

Eronen's self-confessed deflationary account of levels of the organization adds to the interpretation of Monod in the context of Oyama's concerns (Sect. 5.2.2). Most specifically, Eronen's commitment to scale allows us to make sense of the conceptual primacy of genes in ontogenetic and phylogenetic terms. The counterfactual version of Monod's claim can work because it captures the unfolding events in protein synthesis during developmental time scales.[6] Phylogenetically it becomes more complex and may well involve primitive versions of translation and ribosomes in an RNA world prior to the emergence of DNA; those systems then become co-opted and modified into the modern ontogenetic system we observe today (Noller 2012).

Eronen allows us to note the heterarchical interconnectedness of mechanisms and the multiple pathways for data flow within an organism which in turn permit multiple uses for different mechanisms and data, as well as feedback and regulatory effects. To impose an ordered hierarchy at this point would be difficult and largely meaningless. But by looking at local functionality and applying scale considerations, we can develop levels of organization that provide epistemic value and capture key aspects of biology. Thinking clearly about data and information only helps in this task. Oyama's emphasis upon interconnectedness is to note this heterarchy and its causal detail. But within the reticulated organism, we find scale-dependent sequential processes that rely on data for state changes that effectively divert free energy into mechanical work.

[6] Indeed, ontogeny is an events-based ontology of time.

5.3 Robustness

The organisms that genomes are embodied within meet a number of environmental conditions, and at times these conditions will challenge both homeostasis and homeorhesis (the stable trajectory of an organism). Under autopoietic assumptions, we can lay claim to internal processes that will restore balance and direction, and we can further assume that those processes are the result of proximate mechanisms that have an evolutionary history. These mechanisms can be seen as providing *robustness*, enabling the genome to be maintained across variable conditions and thus across generations (Bich 2018). Furthermore, by dint of these mechanisms, organisms can be characterized by their array of available responses to external contingencies, such that they have different ecological bandwidths in terms of the amount of external data they can process. Organisms that can maintain autopoiesis in environments with high levels of spatial and temporal heterogeneity will have higher bandwidth or will be more autopoietic (Gershenson 2015).

The greater the complexity—or changeability—of the ecological niche occupied by an organism then the more data they will need to process in order to maintain themselves, or in everyday Darwinian terms, to survive and reproduce. This means that these organisms are systems that are highly open, ontologically, while retaining their epistemic closure (Luisi 2003). Mayr's discussion of the DNA code fits this view (Mayr 1961, 1984):

> But let us not have an erroneous concept of these codes. It is characteristic of these genetic codes that the programming is only in part rigid. Such phenomena as learning, memory, non-genetic structural modification and regeneration show how "open" these programs are. Yet, even here there is great specificity, for instance with respect to what can be "learned," at what stage in the life cycle "learning" takes place, and how long a memory engram is retained. The program, then, may be in part quite unspecific, and yet the range of possible variation is itself included in the specifications of the code. The codes, therefore, are in some respects highly specific; in other respects they merely specify "reaction norms" or general capacities and potentialities. (Mayr 1961: 1502)

While Mayr does not say it, this view is entirely commensurate with the biological view of information discussed above. At the core of Mayr's comment is the concept of a reactive phenotype. What that means is a phenotype that can vary within parameters, and that variance is in response to something. Thus, Mayr is suggesting that there are broad kinds of traits but that those traits can manifest as quite different tokens. Specific inputs determine specific state changes. His learning example suggests that individuals can learn new associations but that they cannot learn all associations. If learning were simply a case of associating anything with anything, then nothing new would be learned. This is because any given variable would predict all others. This is closely related to the idea of *surprisal*—learned associations need a probability distribution that permits surprise and thus information value.[7] Also, some

[7] Detecting probability distributions and using them to deal with collinearity through data clustering techniques is possibly one aspect of learning (Dickins 2021).

associations can only be learned at certain points in development. He goes on to discuss filial imprinting as an example. This is a form of learning that can occur at a critical period in the development of an organism. In geese, filial imprinting enables a freshly hatched gosling to discriminate a candidate primary caregiver, and the individual that is imprinted upon will be the first organism the gosling sees. In most cases, that will be a parent goose, and so this is an important discrimination to establish early on in what is a precocial bird.[8] Here the program is tightly constrained and enables the development of key discrimination (Bateson 2017). Other social discriminations can occur later in life and throughout life, enabling organisms to maximize key utilities by tracking the politics of their social ecology, for example. More generally, learning can be packaged as a kind of informational autopoiesis, following Cárdenas-García, in which information systems maintain themselves (Cárdenas-García 2020). But we can make a stronger, Mayrian claim that those neurological mechanisms are designed to construct specific, token-level models of particular portions of the external world. They are thus sensitive to kinds of input as a consequence of the DNA code that builds them but can develop quite different models between individuals as a result of experience.

To reemphasize, Mayr's point about DNA codes is that they are open to inputs that will shape phenotypic expression, but the internal conditional logic of that expression is closed. Thus, in keeping with Luisi's comments, they are ontologically open and epistemically closed. What this organization permits are a range of responses to particular kinds of external contingency. Mayr's clear view was that the kinds of contingency and the range of response permitted were a consequence of evolutionary processes, and in particular natural selection (Sect. 4.3.1). Thus evolution affords a certain tolerance toward spatial and temporal heterogeneity (Dickins and Dickins 2008). Moreover, this further implies that the DNA codes are only epistemically closed in the immediate term of their functionality within the individual (barring insult and injury), but they can be reorganized by selection at the population level. This means that natural selection is in the business of developing design solutions for robustness, which means weathering the vicissitudes of life, albeit within parameters. Robustness has a genome-level focus—that is, what is to be preserved—precisely because selection establishes the DNA code, which in turn delimits the proximate mechanisms that process data.

There are a number of different strategies for delivering robustness, and I shall discuss three of them that operate on different time scales: bet hedging and then developmental and physiological plasticity.

[8] Ground nesting precocial birds move around almost immediately upon hatching and so run the risk of isolating themselves from the care that they require.

5.3.1 Bet Hedging

For an organism (and its genome), to survive a number of different conditions that might stochastically present, one solution is to become a generalist and get by in as many environmental states as possible.[9] Compare this to a specialist that can deal optimally with one or a few environmental circumstances. In those circumstances, the specialist will attain higher arithmetic mean fitness than any other kind of strategist; but once conditions shift away from this optimal point, fitness will significantly decline. The specialist will then have to find new places that better suit its specialisms or await a return to optimal conditions at the original site. Waiting for change could obliterate all prior fitness gains. A generalist sacrifices arithmetic mean fitness to get by across various conditions, but gains geometric mean fitness over time, which will enable the selection of generalism (Starrfelt and Kokko 2012).

What is being contrasted here is the notion of a specialist phenotype outcompeting all others and, on average, against the current set of other phenotypes, doing better. When there is little environmental variability, then selection for specialisms should be favored. But where there is variability, then being able to cope (or just do well enough) will enable survival and reproduction, but what is being favored is the multiplicative fitness across generations (geometric mean fitness) which is the mainstay of natural selection—this will enable this trait to remain in the population. The generalist is referred to as a conservative bet-hedger (Haaland et al. 2019).

Diversified bet hedging enables the production of more than one phenotype in a generation. As an example, we might imagine that two different morphs, *a* and *b* can be produced at the point of reproduction. Each morph is specialized to a particular environment, *A* and *B,* respectively. There is a probability distribution for the likelihood of morph *a* or *b* being produced. If the probability of their production matches the encounter rate with *A* and *B*, then the underlying genome will go to fixation. An *a* morph that finds itself in environment *B* will not do as well as a *b* in *B* and vice versa, but at a suitable encounter rate these losses will be buffered. Natural selection will finesse the probability of production in the context of *A* and *B* likelihoods. Diversified bet hedging will outcompete the generalist over the long term, as each diversified phenotype will produce high arithmetic means when in the correct situation. Conservative bet hedging tends to see the reduction of arithmetic mean fitness in favor of geometric means, while diversified bet hedging reduces the fitness correlations between individuals (with the same genotype) (Starrfelt and Kokko 2012). Starrfelt and Kokko see this as a continuum and argue that elements of both approaches can be embodied within singular genotypic strategies (Haaland et al. 2019).

[9]Here I am referring to a local environment changing, thus introducing heterogeneity. I am not considering an organism that moves between environments. For organisms that move to new environments investment in specialization may well become a stable strategy (West-Eberhard 2003).

While bet hedging is a robustness solution, it is random with respect to the environment. In other words, no environmental data is used to induce a particular phenotype during ontogeny. In this sense, the DNA code does not deliver an open program in the sense of a responsive or reactive phenotype. This does not mean that there is no trait variation between individuals, and it is this kind of variation that selection will act upon to stabilize conservative or diversified strategies.

5.3.2 Developmental Plasticity

Developmental plasticity refers to the development of a phenotype in response to data derived from the current environment. In this way, phenotypes are, within limits, fitted to an environment. This robustness strategy is expected when the environmental state at a particular point in development is a good predictor of future environmental states (Meyers and Bull 2002). Meyers and Bull discuss the development of jaw morphology in cichlids. Fish that were fed snails, as opposed to insects, developed significantly larger and more powerful jaw muscles in preparation for that dietary ecology. This case looks like an adaptation, but not all developmental plasticity will necessarily be so, and Meyers and Bull sound a note of empirical caution arguing that each case must be properly inspected.

Nettle and Bateson have directly addressed the issue of adapted developmental plasticity and how to determine it (Nettle and Bateson 2015). Specifically, they focus upon regularities between early exposures and later outcomes, regularities that suggest causation but not necessarily adaptation:

> Field fall crickets whose gravid mothers were exposed to a wolf spider show a heightened immobility response to spider cues, and survive better in an environment containing predatory spiders than crickets whose mothers were not exposed to spider cues. Bulb mites that had rich diets as juveniles develop an aggressive 'fighter' adult phenotype, whereas those that had poorer nutrition as juveniles become non-aggressive 'scramblers'. Starlings that were disadvantaged in competition in the nest go on to fly less well as adults, and women who were separated from their parents before the age of seven show worse health in their forties. (Nettle and Bateson 2015: 1)

Nettle and Bateson claim that the cricket and bulb mite cases seem plausible cases for an adaptationist account, but it is not clear why disadvantaged starlings should fly relatively poorly, or women's health should suffer due to parental separation in childhood. These latter two cases seem mere bad luck and suggest a damage account, but as Nettle and Bateson note, sometimes what seems non-adapted now might have carried advantage in ancestral environments. To address this, they posit a distinction between *informational* and *somatic state-based adaptive*[10] *developmental plasticity* in order to draw out a core principle that unifies them.

---

[10] Bateson and Nettle's use of adaptive is in error I think, but it is a common error in the literature, and one I have made myself. What they really mean is plasticity that delivers phenotypes that are adaptations, where adaptations have a formal definition under standard evolutionary theory, or the

Informational developmental plasticity states that input at a point in development can be used as a cue about the future environment. Evolutionary processes have enabled that input (*I*) to become associated with the adult experience of the future environment (*E*), and a particular phenotype (*P*) is favored in those circumstances (as the authors put it, it is better to have *P* than *not-P*). As a result of selection, it is now the case that exposure to that input will lead to the development of the relevant phenotype as that is more likely to maximize fitness in the predicted future environment. So, what has been established here is a conditional architecture such that $I \rightarrow P$. This is in keeping with the idea that *information* = *data* + *context* (Dickins and Dickins 2008). In this case, *I* is the datum, and the context is *E*. The state change to *P* is relevant in this context. The surprisal value of *I* will be a function of how many possible varieties of environment the organism can meet in its adult state. In the Nettle and Bateson example, the implication is that there are only two possible futures, *E* and *not-E*. The probability distribution of *E* and *not-E* will determine how informative *I* is in mathematical terms.[11] Nettle and Bateson argue that this informational view accounts for cases such as the crickets in the quotation above.

Somatic state-based developmental plasticity is described as follows:

> In such hypotheses, *I* is an input that has a general effect on the developing individual, such that some somatic state variable *S* is enduringly altered. By somatic state variable, we mean a parameter with broad fitness implications such as size, muscular strength, organ capacity or extent of DNA damage. In the example of the bulb mite, *I* is the food available and *S* is size; the less good the food is, the smaller the size the mite can grow to. Importantly, the influence of *I* on *S* cannot be merely transient: a small mite will remain small through adulthood. The other component of somatic state-based models is a phenotypic trait *P* that is potentially independent of *S*. *P* in the bulb mite case is aggression. It would be possible for a small individual to be aggressive, or a large one non-aggressive, so *S* and *P* are potentially independent. However, what has evolved is a mechanism linking the two together: the switching rule 'if you find yourself large, become aggressive; if you find yourself small, become non-aggressive'. For this rule to be adaptive, the optimal phenotype must be dependent on somatic state. That is, expected fitness for a large mite must be higher if it becomes aggressive than if it does not, while for a small mite, the opposite must be true. (Nettle and Bateson 2015: 3)

Nettle and Bateson are keen to distinguish this from an informational account, claiming that the relationship between *I* and *S* is a causal one, such that *S* is a consequence of *I*. This is because they consistently treat *I* as information in the prior

---

Modern Synthesis. More precisely they mean that the reaction norms that permit a range of phenotypes are adaptations. Plasticity connotes flexibility and change, as does the term adaptive. Adaptive can be used at the population level in evolutionary accounts—as in Wright's adaptive landscapes—where the population changes across evolutionary time. But in the current formulation here, adaptive plasticity simply repeats the flexibility claim at the individual level. This becomes a problem because adaptive plasticity permits a view of adaptations as malleable by processes other than selection and drift. A view in stark contrast with Bateson and Nettle's core message.

[11] Shannon information is a logarithmic function. For example, if you flip a coin you have two symbols (heads or tails) that will occur with equal probability. The bits of information per symbol are calculated in this case as $\log_2 2 = 1$. For two coins it would be $\log_4 2 = 2$ and so on.

account and claim that it would be unusual to see a cause as conveying information about its effect in this case. So, for Nettle and Bateson, information has a representational capacity, such that *I* can represent something about a future *E* in their informational plasticity scenarios. But as I have already shown, their informational account makes sense in terms of Shannon type information theory when *I* is treated as a datum. This does not undermine the evolutionary logic of Nettle and Bateson's claim. But it does mean that the somatic state-based scenario can also be seen as informational in the stricter sense, following Floridi, that I have adopted here. This becomes particularly apparent when inspecting the conditional switching rule that they use to account for the emergence of a particular behavioral phenotype, *P*. But I would argue that nutritional inputs can have an informational role in organisms in the context of growth as here. Indeed, only where input-response patterns are chaotic, then we have reason to doubt an informational set-up, but that informational set-up does not have to be an adaptation.

This last point is addressed by Nettle and Bateson, but from their representational take on information, as a unifying principle for determining adapted developmental plasticity:

> (I)ndividuals who experience the developmental input and develop the phenotype must have higher expected fitness than those who experience the developmental input and do not develop the phenotype; whilst individuals who do not experience the developmental input and do not develop the phenotype must have higher expected fitness than those who do not experience the developmental input but do develop the phenotype. (Nettle and Bateson 2015: 4)

Without this patterning, any systematic responding can only be considered informational in either Nettle and Bateson's sense or that of Floridi. It is this patterning that one needs to look for in order to determine whether or not developmental plasticity is an adapted response.

My insistence on reinterpreting Nettle and Bateson with regard to information may seem overly fastidious, after all, I am broadly agreeing with their take on routes to developmental plasticity. But my reason for this is the context of this book. Nettle and Bateson have only set out to make some clear-sighted comments on developmental plasticity with a mind to assisting in determining instances of it. My objective is different. I am seeking to give parameters to organic life and to use Shannon type information to clarify the broad schematics of how organisms are organized. So, I am laying claim to a view of information that will incorporate the functioning of DNA, cells, organs, and so forth. It is a view that thoroughly embeds organisms within environments, can give a measure of their complexity and will help us to understand their relationship to ecological niches.

5.3.3 Physiological Plasticity

Physiological plasticity occurs within individuals across the life span. As with developmental plasticity, this strategy relies on environmental data but to induce physical and behavioral changes such as the ability of organisms to migrate to new environments to access resources, which is a physical response to changes in daylight and temperature modulated by neuro-hormonal changes. Where developmental plasticity calibrates organisms to adult environments, physiological plasticity operates on more immediate time scales, enabling an organism in a direct, responsive mode.

Many physiological processes will maintain dynamic stability and homeostasis but some will also enable compensatory responses to the environment (Bich et al. 2016). Learning as a broad class of activity probably fits both categories.

Dennett has proposed an architectural metaphor, the Tower of Generate and Test (henceforth the Tower) that aims to structure increasing responsive complexity in organisms over evolutionary time (Dennett 1995). The Tower was originally intended to explain the broad phylogenetic changes that led to modern human cognition, but we can adopt it to analyze physiological plasticity. Each floor of the Tower, as ascended through evolutionary time, represents a general change in the ability of creatures to find efficient solutions to contingent problems. Such change leads to an ability to exploit new immediate environments and to maximize average lifetime inclusive fitness. Each floor also schematically represents a typology of extant creatures that inhabit environmental niches that are best suited to that floor's problem-solving technique.

On the first floor, we find Darwinian Creatures, which are generated by natural selection and all possess different hardwired phenotypes. Their solutions to contingent problems are granted by naturally selected adaptations. This kind of design does not preclude bet hedging nor developmentally plastic responses. The kind of informational relationships between inputs and phenotypic outputs described by Nettle and Bateson (2015), while conditional and on a probability distribution are nonetheless the outcome of natural selection.

The second floor is that of the Skinnerian Creatures, who can vary their responses to the environmental contingencies encountered. They possess hardwired reinforcers that bias the creature to what Dennett terms *smart moves*—thus, the creature will vary its response until something good comes of it and will learn it (unless it is a fatal response) for the next time such a contingency is encountered. While this operant learning allows the acquisition of new moves through novel associations (which bring the benefit of certain plasticity when dealing with the environment), classical conditioning or expectancy learning has the advantage of allowing the constant monitoring of the environment. Being able to attune to existing, naturally occurring contingencies, i.e., learning expectancies enables the use of natural indicators and also the ability to detect the unusual. In this way, Skinnerian Creatures can begin to establish informational relationships that are relevant to them by monitoring regularities and enabling surprisal (Cárdenas-García 2020).

Popperian Creatures, on the third floor, can make first moves that are less likely to be fatal due to an ability to pre-select options, so they no longer need to rely upon trial-and-error learning. Popperian Creatures are understood by Dennett in fairly classical cognitive terms and rely upon mental representations of the outer world to calculate what might happen if certain actions were taken. Representations are likely to be stored modularly to meet specific functional needs, but the modules do not speak to one another at all. The Gregorian Creatures, on the fourth floor, go beyond modular encapsulation by having an overarching layer of higher control that is a function of symbolic culture, and Dennett suggests that humans inhabit this floor. Symbolic culture is an external shared memory and enables a sense of *we,* which removes humans from an entirely individualistic perspective (Ginsburg and Jablonka 2019).

Dennett's Tower, at each floor, is a consequence of the generation of variation and the selection of a variant in light of testing against the external world. It is grounded in evolutionary processes understood in standard, Modern Synthetic terms. But upon this foundation rests a series of increasingly plastic types of organism relying upon different architectures all underpinned by neurophysiological mechanisms.[12] These mechanisms are all variants of learning, and what learning does is to fundamentally organize the internal context of the organism so that it responds to particular inputs, either by maintaining its current behavior, readying itself for later action or acting immediately. As Dennett noted, Skinner saw learning as taking over from where hardwired behavioral responses stopped. Operant conditioning could make bespoke use of behavioral responses by shaping the probability of those responses in the context of particular inputs or data, and in this way, produce a reaction norm just as Mayr claimed (Mayr 1961). This does not mean that all learning and all behavior maximize fitness, and the lessons from Nettle and Bateson (2015) regarding the allocation of adapted status apply here also. For example, for any stimulus-response (S-R) pairing to be considered adapted, it must be considered in terms of whether or not the outcome (O) is fitness maximizing (on average). So, it must be the case that the production of R in the presence of S must lead to fitness gains, but the production of R in the presence of not-S should not.

The Gregorian strategy relies on the communicative aspect of symbolic language, but to effectively communicate and thus cooperate as a *we,* organisms must be able to deal with signal ambiguity. In linguistics, this was traditionally the area of pragmatics, which dealt with the ability to draw inferences about the intended interpretation of an utterance. Ambiguity amounts to an input being relevant in more than one context, leaving the recipient of the input to determine which context was intended by the actor. Relevance theory (RT) has provided a theoretical account of the process:

> (R)elevance... is defined as a tradeoff between two competing properties of an utterance. On the one hand, there are the worthwhile changes in the receiver's representation of the world that are warranted by the utterance. These are called positive cognitive effects and include the strengthening, weakening, or elimination of previously held knowledge and the

---

[12]For example, we can start simply with the Hebb rule (Hebb 1949) which states that if a synapse becomes repeatedly active at about the same time that the postsynaptic neuron fires, changes will take place in the structure or chemistry of the synapse that will strengthen it, facilitating stimulus-response pairings. More generally, his theory of cell assemblies was fundamentally behaviorist in scope (Huyck and Passmore 2013).

> provision of premises from which to infer new knowledge. For example, if a colleague's utterance leads me to a better understanding of some problem, by any of these means, then it has had a positive cognitive effect upon me. As such, positive cognitive effects can be seen as the payoff associated with correct utterance interpretation. To achieve these cognitive effects, some time and energy must be expended. This processing effort must be weighed against the cognitive effects of an utterance, on the other hand. The result gives us a measure of relevance... Thus, all other things being equal, the greater the positive cognitive effects achieved by processing an utterance, and the lower the processing effort expended in processing the utterance, the greater the relevance of that utterance. This is how relevance is defined within RT. This means, of course, that the same utterance will have different degrees of relevance to different individuals and at different times: The listener's prior knowledge will impact on both the degree of positive contextual effect achieved by an utterance and the amount of processing effort required to comprehend it. (Scott-Phillips 2010: 586)

The idea that an utterance, as an input, can produce changes in a cognitive system that might be beneficial is in keeping with Shannon type information. But we should note that the cognitive system is portrayed as generating a representation of the world that can be updated and is something that is probably created by the recipient of the utterance. This fits with broad learning accounts, and we do not need to commit to a particular view on representations or models at this point (Butlin 2020). To gain maximum relevance, the positive benefits of an utterance must outweigh the costs of the efforts to apply the utterance to existing structures of knowledge through various inferential processes. Each update of a model due to an incoming utterance is not assumed to have a fitness beneficial effect, but overall relevance finding enables greater social coordination, and this will have downstream fitness benefits (Scott-Phillips 2010). This kind of situation is interesting because new inputs can cause a state change that is effectively an update of the context for any further inputs. In this way, the informational value of inputs is altered through time, which is a characteristic of ongoing learning (Boisot and Canals 2004). In this way, learning generally understood is an open system both ontologically and to a limited extent epistemically. Those limits will be due to path dependency in learning, as a consequence of prior modelling and because learning is fundamentally about prediction (Clark 2013; Ginsburg and Jablonka 2019).

Organisms on higher floors of the Tower will be endowed with those capacities that went before, meaning that their ability to coordinate highly flexible responses to the environment will be increased. Thus, while I have established bet hedging and developmental plasticity as distinct robustness strategies, it is entirely possible that there is interaction. For example, Frankenhuis and colleagues have made a case for diversified bet hedging that leads to offspring morphs that are more or less developmentally plastic (Frankenhuis et al. 2016). This can be interpreted as more or less sensitive to incoming environmental data, implying that there will be some costs to plasticity under certain environmental conditions, and it should not always be switched on. And just as they have suggested a relationship between bet hedging and developmental plasticity, it is entirely possible that developmental responses will fine-tune physiological responsiveness to environmental parameters and facilitate learning (Dickins and Rahman 2012; Frankenhuis et al. 2018).

Learning, and more generally physiological plasticity, is fundamentally about trying to reduce surprisal by organizing internal context such that inputs are expected, and the system is ready for it. This is working to reduce the informational value of inputs by changing the system—and can be seen as homeostasis in keeping with autopoietic views. As the world constantly changes, inputs continue to have informational value. But this last is only possible because of the array of responses made available to the organism as a result of selection. Thus, the fundamental Darwinian Creature lays the groundwork for the greater flexibility of the Skinnerian, Popperian, and Gregorian creatures.

5.4 Summary and Conclusion

Dennett's Tower is a hierarchy of levels. Some creatures are purely Darwinian, others Darwinian and Skinnerian, and so on. But this should not be taken to imply that Eronen's deflationary account of levels is abandoned (Eronen 2015). Dennett's intuition pump is designed to make us think about phylogenetic transitions, and I have deployed it in the context of a discussion about plasticity because ascending the Tower marks a transition toward greater flexibility. All of this flexibility will be achieved by a heterarchical network of data dependent mechanisms as discussed above. But greater flexibility does not necessarily mean greater complexity or even greater informational potential, for it is perfectly possible to have a Darwinian creature with a large genome and a complex bet hedging structure that yields impressive robustness. However, in general terms, we can understand the addition of developmental and physiological plasticity as additions of informational complexity to the ancestral design. It is informational complexity because selection operates over the relationship between data and context. Those systems, or contexts that make good use of data thrive.

The view of information espoused here clearly fits with the idea of design that has been at the heart of natural theological and evolutionary considerations. The mechanisms of organisms appear prepared for the inputs they receive due to the constancy of their response and also the complexity of its delivery. This speaks to the language of adaptation, as clarified by Williams (1996). Given this, the critic may well brush my analysis to one side, claiming that it is necessarily partisan and merely redescribes a theory that they disagree (or agree) with for independent reasons. If I were attempting to use this view of information to support the foundational principles of the Modern Synthesis, this would be a fair criticism. But this is not my intention. Instead, this analysis, this perspective on information is given in an attempt to clarify the nature of the Modern Synthesis, the kind of explanation it is actually delivering.

Information, as a concept, has been deployed many times in discussion of evolution and in biology more generally, but it is commonly the case that its use has been loose and metaphorical. When information is regarded as something that is harvested and something that is a complete, representational instruction, there is the potential for secondary assumptions to hitch a ride. For example, when this interpretation is made during discussions of temporal, causal sequence and necessary conditions for protein synthesis, then gene-level theory is readily interpreted as a preformationist idea in much the way Oyama (2000) has. And she is in good company. My plea is for clarity when using this and related concepts and charity when reading those who deploy informational terms.

There is more to my focus upon information as *data + context* than simply trying to understand programmatic statements from leading biologists. I also maintain that the job of biology is one of uncovering informational relationships in the natural world, of determining what is relevant in the environment, by studying the responses of organisms, and their sub-systems, to inputs. More strongly, I think that the emergence of Darwinian thought and its refinement through the Modern Synthesis was a realization of this fundamental relationship. The human quest for meaning is the business of understanding the effects of environmental exposures throughout life. The poet captures this through descriptions of her phenomenological experience but however rich her description, they are invariably divorced from the relevant inputs that caused her reverie.[13] As an artist, she tries to capture the entirety of the relationship between inputs and person by focusing solely upon the outcome. What is biologically meaningful is that which responds to data, which is context and contexts are provided by systems. Evolutionary biology gives an account of why the systems we see persist and what is used as data in the world.

In the final chapters, I shall put this view of information to work in dealing with critics of the Modern Synthesis. In keeping with Oyama's comment on Monod, when I read the detail of the processes that these critics mobilize, I can see they understand biology in a way compatible with the view outlined in this chapter. But, when I read their programmatic statements about the inadequacies of the Modern Synthesis, I am, well, surprised.

References

Bateson P (2017) Behaviour, development and evolution. In Behaviour, Development and Evolution. Open Book Publishers. https://doi.org/10.11647/obp.0097

Bateson P, Gluckman P (2011) Plasticity, robustness, development and evolutin. Cambridge University Press

Bechtel W, Bich L (2021) Grounding cognition: heterarchical control mechanisms in biology. Philos Trans R Soc B Biol Sci 376(1820):20190751. https://doi.org/10.1098/rstb.2019.0751

Bich L (2018) Robustness and autonomy in biological systems: how regulatory mechanisms enable functional integration, complexity and minimal cognition through the action of second-order control constraints. In: Bertolaso M, Caianiello S, Serrelli E (eds) Biological robustness: emerging perspectives from within the life sciences. Springer, New York, pp 123–147. https://doi.org/10.1007/978-3-030-01198-7_6

[13] I suspect a key pleasure of more abstracted arts is the effort of trying to marry a phenomenological expression with an input. You read a poem and the sensation of it clicking, of it making sense for you, occurs when you can determine a context and an input in which that state would make sense. Then the poem acts more like a simulation. Dennett would see this as a Gregorian attribute.

Bich L, Mossio M, Ruiz-Mirazo K, Moreno A (2016) Biological regulation: controlling the system from within. Biol Philos 31(2):237–265. https://doi.org/10.1007/s10539-015-9497-8

Boisot M, Canals A (2004) Data, information and knowledge: Have we got it right? J Evol Econ 14 (1):43–67. https://doi.org/10.1007/s00191-003-0181-9

Butlin P (2020) Cognitive models are distinguished by content, not format. Philos Sci 88 (January):83–102. https://doi.org/10.1086/710030

Cárdenas-García JF (2020) The process of info-autopoiesis – the source of all information. Biosemiotics 13(2):199–221. https://doi.org/10.1007/s12304-020-09384-x

Clark A (2013) Whatever next? Predictive brains, situated agents, and the future of cognitive science. Behav Brain Sci 36(3):181–253. https://doi.org/10.1017/S0140525X12002221

Clark A (2018) A nice surprise? Predictive processing and the active pursuit of novelty. Phenomenol Cogn Sci 17(3):521–534. https://doi.org/10.1007/s11097-017-9525-z

Coall DA, Callan AC, Dickins TE, Chisholm JS (2014) Evolution and prenatal development: an evolutionary perspective. In: Lamb M (ed) Handbook of child psychology and developmental science: Vol. III. Cambridge University Press, Cambridge, pp 57–105

Cohen IR (2000) Tending Adam's garden: evolving the cognitive immune self. Academic Press

Craver CF, Bechtel W (2007) Top-down causation without top-down causes. Biol Philos 22 (4):547–563. https://doi.org/10.1007/s10539-006-9028-8

Dawkins R (1989) The selfish gene, 2nd edn. Oxford University Press, Oxford

Dennett DC (1995) Darwin's dangerous idea: evolution and the meanings of life. The Penguin Press, Allen Lane

Dickins TE (2020) Conflation and refutation: Book Review of Uller, T. and K. N. Laland. eds. 2019. Evolutionary causation: biological and philosophical reflections. MIT Press, Cambridge, MA. pp. ISBN: 978-0-262-03992-5. $60.00/£50.00. Evolution, 508–514. https://doi.org/10.1111/evo.13916

Dickins TE (2021) Ecological psychology, radical enactivism and behavior: an evolutionary perspective. Behav Philos 49:1–23

Dickins TE, Dickins BJA (2008) Mother Nature's tolerant ways: why non-genetic inheritance has nothing to do with evolution. New Ideas Psychol 26(1):41–54. https://doi.org/10.1016/j.newideapsych.2007.03.004

Dickins TE, Rahman Q (2012) The extended evolutionary synthesis and the role of soft inheritance in evolution. Proc R Soc B Biol Sci 279(1740):2913–2921. https://doi.org/10.1098/rspb.2012.0273

Dretske FI (1983) Précis of Dretske of knowledge and the flow of information. Behav Brain Sci 6 (1):55–63

Eronen MI (2015) Levels of organization: a deflationary account. Biol Philos 30(1):39–58. https://doi.org/10.1007/s10539-014-9461-z

Floridi L (2003) From data to semantic information. Entropy 5(2):125–145. https://doi.org/10.3390/e5020125

Floridi L (2010) Information: a very short introduction. Oxford University Press, Oxford

Frankenhuis WE, Panchanathan K, Barto AG (2018) Enriching behavioral ecology with reinforcement learning methods. Behav Process 161(February 2018):94–100. https://doi.org/10.1016/j.beproc.2018.01.008

Frankenhuis WE, Panchanathan K, Belsky J (2016) A mathematical model of the evolution of individual differences in developmental plasticity arising through parental bet-hedging. Dev Sci 19(2):251–274. https://doi.org/10.1111/desc.12309

Gershenson C (2015) Requisite variety, autopoiesis, and self-organization. Kybernetes 44 (6–7):866–873. https://doi.org/10.1108/K-01-2015-0001

Ginsburg S, Jablonka E (2019) The evolution of the sensitive soul: learning and the origins of consciousness. MIT Press, London

Godfrey-Smith P (1996) Complexity and the function of mind in nature. Cambridge University Press

Godfrey-Smith P (2007) Information in biology. The Cambridge Companion to the Philosophy of Biology, 103–119. https://doi.org/10.1017/CCOL9780521851282.006
Godfrey-Smith P (2017) Complexity revisited. Biol Philos 32(3):467–479. https://doi.org/10.1007/s10539-017-9569-z
Griesemer J (2015) The enduring value of Gánti's chemoton model and life criteria: Heuristic pursuit of exact theoretical biology. J Theor Biol 381:23–28. https://doi.org/10.1016/j.jtbi.2015.05.016
Haaland TR, Wright J, Tufto J, Ratikainen II (2019) Short-term insurance versus long-term bet-hedging strategies as adaptations to variable environments. Evolution 73(2):145–157. https://doi.org/10.1111/evo.13659
Hebb DO (1949) Organization of behavior: a neuropsychological theory. Wiley.
Henkin TM, Yanofsky C (2002) Regulation by transcription attenuation in bacteria: How RNA provides instructions for transcription termination/antitermination decisions. BioEssays 24 (8):700–707. https://doi.org/10.1002/bies.10125
Huyck CR, Passmore PJ (2013) A review of cell assemblies. Biol Cybern 107(3):263–288. https://doi.org/10.1007/s00422-013-0555-5
Luisi PL (2003) Autopoiesis: a review and a reappraisal. Naturwissenschaften 90(2):49–59. https://doi.org/10.1007/s00114-002-0389-9
Maynard-Smith J, Szathmary E (1995) The major transitions in evolution. Oxford University Press, Oxford
Mayr E (1961) Cause and effect in biology. Science 134:1501–1506. https://www.ifor.math.ethz.ch/education/bachelor/lectures/fs2011/math/Biomath/Mayr.pdf
Mayr E (1984) The triumph of evolutionary synthesis. Times Literary Supplement, November(2), 1261–1262
Meyers LA, Bull JJ (2002) Fighting change with change: adaptive variation in an uncertain world. Trends Ecol Evol 17(12):551–557. http://dialnet.unirioja.es/servlet/articulo?codigo=2851914%5Cnpapers://994295af-42ed-47fe-9a2a-07e2e7d6a6b8/Paper/p709
Nettle D, Bateson M (2015) Adaptive developmental plasticity: what is it, how can we recognize it and when can it evolve? Proc R Soc B, 282(1812), 20151005. https://doi.org/10.1098/rspb.2015.1005
Noble D (2011) Neo-Darwinism, the modern synthesis and selfish genes: are they of use in physiology? J Physiol 589(Pt 5):1007–1015. https://doi.org/10.1113/jphysiol.2010.201384
Noller HF (2012) Evolution of protein synthesis from an RNAworld. Cold Spring Harb Perspect Biol 4(4). https://doi.org/10.1101/cshperspect.a003681
Oppenheim P, Putnam H (1958) Unity of science as a working hypothesis. In Minnesota studies in the philosophy of science (Vol. 2, pp. 3–36)
Oyama S (2000) The ontogeny of information: developmental systems and evolution, 2nd edn. Duke University Press
Popper KR (1945) The open society and its enemies. Routledge
Queller DC, Strassmann JE (2009) Beyond society: the evolution of organismality. Philos Trans R Soc Lond Ser B Biol Sci 364(1533):3143–3155. https://doi.org/10.1098/rstb.2009.0095
Scott-Phillips TC (2010) The evolution of relevance. Cogn Sci 34(4):583–601. https://doi.org/10.1111/j.1551-6709.2010.01095.x
Shannon CE (1948) A mathematical theory of communication. Bell Syst Tech J 27(3):379–423
Starrfelt J, Kokko H (2012) Bet-hedging-a triple trade-off between means, variances and correlations. Biol Rev 87(3):742–755. https://doi.org/10.1111/j.1469-185X.2012.00225.x
Sultan SE (2019) Genotype-environment interaction and the unscripted reaction norm. In: Uller T, Laland KN (eds) Evolutionary causation: biological and philosophical reflections. MIT Press, London, pp 109–126
West-Eberhard MJ (2003) Developmental plasticity and evolution. Oxford University Press, Oxford
Williams GC (1996) Adaptation and natual selection: a critique of some current evolutionary thought, 2nd edn. Princeton University Press

Evolution and Development

6

Abstract

Evolutionary developmental biology, or evo-devo, has emerged as a subdiscipline of biology and as a challenger to Modern Synthesis. In this chapter, I briefly review its emergence and then move to discuss key positions from within evo-devo on phenotypic novelty and on the interpretation of reaction norms, both of which have formed the basis for criticism of the Modern Synthetic view of development. I argue that evo-devo has misunderstood the view of development from the Modern Synthesis, such that developmental accounts from within evo-devo are well suited to the parameters imposed by the black boxing of development during the synthesis. I also discuss how colloquial uses of information have informed an erroneous view of the role of genes, a view readily corrected by adopting a more formal view of *information = data + context.*

Keywords

Development · Form · Adaptive change · Novelty · Facilitated variation · Weak linkage · Exploratory processes · Compartmentation · Genetic regulation · Reaction norms · Developmental program · Information · Data · Context

6.1 Introduction

Modern work on the role of development in evolution addresses the most fundamental question of evolutionary biology: what is evolution? Or, put another way, what is evolutionary theory a theory of?

Is it a theory of the diversity of phenotypes that we see in the world? So, an account of why there are so many different forms as well as an account of their distribution?

Or is it a theory of the production of phenotypes, of the creation of form?

T. E. Dickins, *The Modern Synthesis*, Evolutionary Biology – New Perspectives on Its Development 4, https://doi.org/10.1007/978-3-030-86422-4_6

In Chap. 1, I claimed that many of those arguing for an extension to the Modern Synthesis are approaching the discipline with a focus upon *form*. Indeed, as I noted in that chapter Pigliucci has tied extension to the need for a mechanistic theory of form, in contrast to the current evolutionary focus upon genes, or as I argue in Chap. 1, upon populations (Pigliucci 2007). This, then, is a fundamental tension between the Modern Synthesis and any Extended Evolutionary Synthesis.

Much of the argument in favor of extension is about the role of natural selection and the concept of adaptation. It is proposed that developmental processes are capable of producing adaptations, independently from natural selection, and thus that they should be regarded as components of evolutionary change. Following from this, advocates of extension challenge what they see as the primacy of natural selection that emerged during the hardening phase of the Modern Synthesis (Gould 2002). The claimed acceptance of natural selection as the predominant creative force in evolution is thought to have discounted any role for development.

In this chapter, I will argue that this debate is prone to the conflation between proximate and ultimate causation, and transformational and variational views of evolution, that I discussed in Chap. 4. Darwinian evolution moved science away from transformational views to population thinking. This, along with natural selection, was firmly incorporated within the Modern Synthesis, and thus the synthesis was not about the creation of form but rather the sorting of it. However, the focus upon Darwinian adaptations and the emergence of a gene-based understanding of what an adaptation was in selection terms meant that the diversity of specific forms seen in populations were claimed as part of the explanatory target of standard evolutionary theory. This did not mean that the architects of the synthesis felt that they could account for the production of form, but rather the amount and kind of its diversity. However, discussions about adaptations and their underlying genes have been interpreted as an attempt to explain the production of form, such that selection yields the requisite genes to build adaptations. This inference has been facilitated by Mayr's discussions of DNA codes and programs (Chap. 4), which some have understood in a deterministic and form-directed vein. Thus, the idea has emerged that the Modern Synthetic view is that only genes are required to build an adaptation, and form is predominantly adaptation achieved via natural selection.

One of the architects of the Modern Synthesis had this to say about form:

> The evolution of complex organisms rests on the attainment of gene combinations which determine a varied repertoire of adaptive cell responses in relation to external conditions. The older writers on evolution were often staggered by the seeming necessity of accounting for the evolution of fine details of an adaptive nature, for example, the fine structure of all of the bones. From the view that structure is never inherited as such, but merely types of adaptive cell behavior which lead to particular structures under particular conditions, the difficulty to a considerable extent disappears. (Wright 1931: 147)

While this statement is firmly embedded within a view of genetic inheritance, it is also a view that requires developmental processes to deliver adapted form. Indeed, it is a claim that genes deliver developmental capacities that then take their cue from the external world. This view is, I think, the Modern Synthetic view of development, and it implies that developmental capacities are under genetic influence and constraint. But this in no way undervalues the role of development in biology and nor does it imply that development cannot affect evolutionary outcomes.

The chapter will begin with a discussion of where development sits in the Modern Synthesis and then move to some key arguments from developmental perspectives among those advocating for an extension. These arguments will be critically evaluated, with reference to the concept of information introduced in Chap. 5, before moving to some concluding comments.

6.2 The Place of Development in the Modern Synthesis

Mayr was clear in a number of publications that development was a separate process from that of evolution. The former was something that happened within individuals, the latter to populations. But he was equally clear that developmental biologists were not actively excluded from the Modern Synthesis but simply chose not to attend (Chap. 4; Mayr 1984, 1992). That he thought the two processes were separable does not entail that he felt there was no interaction between the two. And the fact that developmentalists were invited to join the key meetings of the Modern Synthesis in which he was involved suggests those engineering the synthesis felt there would be fruitful science to be done between the two kinds of scholars. Nonetheless, Mayr is often cited as an obstacle to the integration of development with evolutionary theory (Laland et al. 2011; Scholl and Pigliucci 2015).

Those making the claim that development was blocked from incorporation into the Modern Synthesis see a number of roles for evolution. Rather than focus upon the process of development, the variation introduced by that process becomes a source of novelty for selection to operate over. This variation can be understood in terms of the effects of regulatory genes on phenotypic outcomes. Others see development as a cause of adaptation and a route to the spread of that adaptation through a population, as we see in niche construction theory (Depew 2017; Laland et al. 2011). These ideas are fully part of what is now referred to as evolutionary developmental biology (or evo-devo).

Evo-devo is in many ways an attempt at synthesis between evolutionary and developmental biology focused upon understanding the emergence of the phenotype, and it is sometimes historically linked to recapitulation arguments of the sort made by Haeckel (Gould 1977; Müller 2007). Müller argues that it in fact has multiple historical routes but:

> (T)he fundamental motive force underlying its formation was not the reactivation of earlier research goals, nor a single technical discovery, but the increasing awareness of important explanatory deficits of the leading paradigm of evolutionary biology, adaptation. Criticism of adaptationism and of the powers of natural selection is old, but in the late 1970s and early 1980s, concern accumulated about the inability of neo-Darwinian theory to account for a number of characteristic phenomena of phenotypic evolution. These included the biases in the variation of morphological traits, rapid changes of form evident from the fossil record, the origin of nonadaptive traits, apparent dissociations between genetic and phenotypic

> evolution, and the origination of higher-level morphological organization such as homology, body plans, and novelty—to name but some of the major open questions. (Müller 2007: 500)

Müller claims that the cause of these failures was the bracketing, or black boxing, of development and thus an absence of rules for generating a phenotype from a genotype. A number of theorists in recent years have emphasized a need to explain the nature and origin of the form (structure or morphology), and their concern is that standard evolutionary theory has predominantly become a theory of genes, where once it was designed to account for form (Laland et al. 2015; Love 2017; M. Pigliucci 2007).[1] Müller states that evo-devo has become a distinct discipline addressing this problem in two parts: the influence of evolution on development and the influence of development on evolution. He sees evo-devo as "an integrative, profoundly organismal program addressing the causal and reciprocal interrelations between development and evolution at multiple scales and multiple levels of analysis" (Müller 2007: 505). Its conceptual focus is upon what Müller terms the problem of *phenotypic evolution*.

6.2.1 Adaptive and Adapted Views

What does *phenotypic evolution* mean? Under one interpretation, Müller is simply telling us that development is patterned and that patterning is possibly a consequence of selection. Moreover, development as a plastic response enables particular phenotypes to emerge under certain data regimes. If those regimes are biased in a particular direction, this might change selection dynamics for the population. This seems uncontroversial and in accord with Modern Synthetic views, and especially Mayr's view that selection operates over the phenotype. Thus, phenotypic evolution may simply refer to the idea that populations can change in their mean phenotypic traits over time.

It is possible, however, that Müller is hinting at a transformation interpretation, that phenotypic change, within the individual during development, is an evolutionary process. In this way, development and evolution are to be seen as micro- and macro-level versions of the same process, and the approach to evolution is as an aggregate explanation rather than a variational one (Sober 1984). This is in direct contrast with the central view of evolution within the Modern Synthesis: *evolution is a kind of change that happens to populations, not to phenotypes*. Phenotypes can be selected and change their proportional representation, but the source of their intrinsic change lies within the individual. This was Mayr's point about the separation of developmental and evolutionary processes. This is not peculiar to the Modern Synthesis but was in fact a key innovation from Darwin, who shifted evolutionary thinking away from the transformation associated with Lamarck (Chap. 2).

[1] It should be noted that Mayr was committed to the view that natural selection operated over phenotypes, and not genotypes, and that those phenotypes could vary due to developmental processes (Mayr 2002).

Proximately caused variation is a key component of natural selection which enables population-level change, in combination with differential success and inheritance.

Depew discusses these two options with historical reference to Spencer in the context of an analysis of adaptation (Depew 2017). Accordingly, Spencer did not see adaptations as naturally selected *for* but rather as the outcome of dynamic and developmental adjustments of the whole. Here we see the distinction between *adaptive* and *adapted*, a distinction between individual plasticity and a naturally selected trait (Chap. 1). These two categories of things are not mutually exclusive, but nor are they identical. Indeed, these Spencerian adjustments were not unknown to Darwin, according to Depew, but he did not see them as adaptations even when he later claimed a core role for the environmental modification of species.

Depew gives the example of whiteness in arctic creatures. This could be caused by physical and physiological effects, say in response to the cold. These effects might mask underlying trait variation and remove an opportunity for natural selection. Such whiteness may convey an advantage (crypsis), but this in itself is not evidence of adaptation as an evolutionary connection between cause (physical factors) and effect (crypsis) has not been made. Depew's point is that for Darwin this link was essential for terming a trait an adaptation, but for Spencer this was not the case. He would incorporate these single generational, environmentally induced developmental effects into an overarching story that included transgenerational effects of the sort Darwin favored. Darwin's theory was that of gradual natural selection operating over a long time and leading to distinct population characteristics that solved contingent problems. Darwin understood that if this axiom were falsified, all else would be unsound.

The difference between Spencer and Darwin touches directly upon discussions of the proximate–ultimate distinction (Chap. 4, Mayr 1961). Mayr would simply determine that Spencer had conflated proximate environmentally induced development processes with the ultimate process of natural selection. Depew's analysis of this conflation is that Spencer operated in a *selection against* mode, focused at the single generation level, while Darwin operated in a *selection for* mode at the transgenerational level. He argues that the Spencerian and Darwinian conflation is captured by Spencer's phrase "the survival of the fittest," which Darwin latterly adopted himself. The result: ". . . people who called themselves Darwinians in, say, 1880, ascribed adaptedness or fitness to the responsiveness of organisms to environmental challenges independently of the workings of natural selection, which they took to be a pruning process" (Depew 2017: 41). But Depew claims this changed again in the 1890s and throughout the Modern Synthesis.

Depew discusses Weismann, the emergence of neo-Darwinism (Sect. 3.3.1) and the central claim that adapted traits emerge only from natural selection operating over germline variation. This brought the explicit focus upon hard inheritance leading to statistical, population-level accounts of natural selection (Sect. 3.3.2). He is careful to note that Weismann, as an embryologist, saw the role of ontogeny in the emergence of organisms, but Weismann had natural selection operating early in the development chain. This meant that he recognized all kinds of subsequent ontogenetic processes and took a fairly holistic view of the emergence of the adult phenotype. Nonetheless, other theorists such as Morgan bracketed embryology and germline selection to pursue Weismann's genetic questions, regarding this bracketing as a temporary expedient while constructing a portion of the overall account. It is this bracketing strategy that Depew, as well as Müller (Müller 2007) claim continued through the resolution of the biometric and Mendelian debates and was core to the construction of population thinking and the basis of the Modern Synthesis. Although not all theorists involved in the Modern Synthesis were uninterested in development. For example, Depew quotes Dunn and Dobzhansky's 1946 book, *Heredity, Race and Society*: "What is inherited is not this or that trait, but the manner in which the organism responds to the environment" (Depew 2017: 45), which clearly follows Wright's line of argument quoted above (Sect. 6.1, Wright 1931). Under this view, organisms are developmental systems, which is in keeping with contemporary views on developmental plasticity and robustness (Chaps. 5 and 9).

Depew's analysis does not tell us how to interpret Müller's interest in phenotypic evolution, but it establishes two possible interpretations and historically grounds them for us. Moreover, it makes clear one view on what the Modern Synthesis did, and that was to bracket development. In doing this, development was firmly situated as a process to account for within evolutionary biology, but not as an aspect of evolutionary theory. A key reason for this, which has been much emphasized in this book, is the distinction between individual and population-level processes and accounts. But another may also be to with a deeper commitment to scales of process, in keeping with a view on levels of biological organization (Sect. 5.2.1, Eronen 2015). In other words, there is a specific ontological commitment about how organisms are organized and how that organization has its effects.

6.3 Aspects of Evo-devo

> By the mid-twentieth century ... a consensus view of evolution, sometimes called the Modern Synthesis, had incorporated population genetics, selection, and chromosomal inheritance into a robust model of evolution. What was still missing were the cellular and molecular mechanisms underlying the generation of the phenotype, particularly the anatomy, physiology, and behavior of multicellular organisms. How relevant such mechanistic understandings were to the theory of evolution was then unclear. (Kirschner and Gerhart 2010: 253)

Kirschner and Gerhart declare that evolutionary theory would be incomplete without an understanding of phenotypic variation (2010: 253). But they hope to use modern understanding of this variation to show that natural selection is not the "sole creative force in evolution" (2010: 254).

The emphasis upon natural selection as a creative force, and specifically as either the principal or sole force, is common in evo-devo literature. There is an assumption that this is what the Modern Synthesis claimed based upon Gould's analysis of the second, hardening phase of the synthesis (Gould 2002). But such criticisms are never entirely explicit about what is meant by creativity. If the creativity under discussion is that of establishing populations of a certain sort, then natural selection has a distinct role in that, according to the Modern Synthesis. But if creativity is the production of variation in phenotypes within a population, then, while natural selection under some conceptions has *permitted* this array, that permission is only at the level of the conditional architectures of development. The patterns of decisions made during ontogeny have their own causes.

The second form of creativity—the creation of the phenotype—must include the arrival of novel phenotypes which demand some standard of novelty (Moczek 2019). Moczek cites Müller and Wagner's definition in which a novel trait is that which is not homologous to any other in ancestral organisms nor homonomous to any other trait in the same organism (Moczek 2019: 65) but then goes onto note that traditional models of homology are wanting because they have ignored evo-devo. What many evo-devo scholars are keen to claim, including Moczek, is that any novelty, indeed any causal model of the phenotype, cannot rest solely upon genes and genetic variation. This view is often related to claims that the Modern Synthesis is gene centric (Depew 2017; Noble 2011; Otsuka 2019) and refers directly to the crystallized replicator-vehicle view of evolutionary theory given by Dawkins in *The Selfish Gene* ((Dawkins 1976), but see also (Hull 1980)). This is readily interpreted as ignoring development, but as discussed (Sect. 3.4), the selfish gene concept is a heuristic enabling a certain kind of explanatory activity. Dawkins in no way ignored the role of development in the production of the phenotype and directly discussed genes as catalysts in a long-chain developmental process. But, in keeping with Mayr and others, bracketing developmental processes enabled an idealization of complex causality, which delivered useful predictive and explanatory leverage when modelling whole organism evolutionary dynamics.

Kirschner and Gerhart (2010) are interested in novelty which they say "arises from an interplay between the properties of the organism and mutation under selection. The nature of the developmental and cellular circuits contributes a great deal to the kinds of variation that selection can act upon." (2010, 254).

> (P)henotypic variation is assumed to be well behaved in coupling genotype and selection. But its role in evolution appears not to be limiting if indeed, as is often assumed, it is small in its extent of change, it is copious in its amount, and it is isotropic in direction – that is, of many, perhaps all, kinds yet with no relation to the selective environment. (Kirschner and Gerhart 2010: 255)

Here then is their problem. They think that phenotypic variation is unruly, and its mapping to the underlying genotype is problematic, which makes the underlying genotype potentially invisible to the selection, or at least partially occluded. Moreover, they think that the standard models simply assume that genotypic change leads to phenotypic change without explaining how. This is another version of the bracketing complaint—developmental processes have been black boxed to deliver neat quantitative population genetic models, or inclusive fitness models assuming the phenotypic gambit (Sect. 3.4, Grafen 1984). This means that form has been removed in that no one can say how "a zebra was actually formed in evolution." (Kirschner and Gerhart 2010: 255).

In making this claim, Kirschner and Gerhart are simply telling us that the Modern Synthesis did not seek to explain form but rather population change. They spend a little time asking possible questions that a focus on form might allow—for example, what is modified during descent with modification, and how many genetic changes are needed for a modification? In asking these questions, it becomes apparent that they are focused upon the proximate questions of development and also that they understand that developmental trajectories are underpinned by genes. Furthermore, they are not (yet at least) claiming the Modern Synthesis saw biology through a one gene-one trait lens but understood there was a huge complexity with regard to development. Kirschner and Gerhart are really seeking generalities about phenotypic emergence, and they build a theory to do this.

Their theory building relies on three parts:

1. A discussion of the deep conservation of process at the cellular level, such that these conserved processes are repeatedly used in the generation of the phenotype.
2. A discussion of the general properties of biological systems that deconstrain phenotypic change.
3. An argument that physiological variation is an important substrate for evolutionary change.

Kirschner and Gerhart's core claim is that phenotypic novelty

> achieved in the course of variation and selection... may contain rather little that is new. In most cases the components are largely unchanged, and the novelty rests on regulatory changes, such as in moving the expression of a genetic program in time, or stabilizing and enhancing what was already present as a physiological or developmental adaptation. If true, this is profoundly important, as it greatly reduces the expectations about what genetic change has to accomplish. (Kirschner and Gerhart 2010: 259)

The relationship between the phenotype and the genotype no longer appears unruly, but rather conservative.

6.3.1 Conserved Genes and Core Processes for Evolution

Kirschner and Gerhart (2010) note that we share 50% of our 21,000 genes with Drosophila, and our genome is only around 50% larger. This is an example of the conservation of protein structure and function across large phylogenetic spans. Conservation is at the heart of their developmental claims.

They begin their analysis by inspecting a claim made by Mayr in his 1963 book *Animal Species and Evolution*. Accordingly, Mayr saw no role for homology in accounting for similarities between structures such as bat and bird wings, but instead felt they were due to convergence, the independent emergence of similar features in different species where a related feature was not present in their last common ancestor. By denying a role for homology Mayr is effectively claiming each wing as a novelty, caused by separated instances of genetic variation in each lineage and then selection fixing the emergent traits within the population. This view is credited to a lack of knowledge at that point in the twentieth century. Two decades later, homology is being invoked in arguments that presaged the arrival of evo-devo where a central criterion for determining two traits homologous was the establishment of common developmental pathways (Roth 1984).

> Implicit in this discussion is that homology is not an all-or-none phenomenon: it is important to recognize degrees of homology. The hierarchical nature of phylogenetic homology is widely recognized: as forelimbs, bird wing and bat wing are homologues; as wings they are not – the reptilian common ancestor had forelimbs, not wings. By analogy with phylogenetic relationship, in which taxa can be closely or distantly related, structures can be biologically homologous to varying degrees – from nearly identical, strongly homologous structures, on down to the very weakest degree of homology, manifested by structures which simply derive from the same germ layer. The points at which developmental paths diverge determine the strength of the homology. (Roth 1984: 18–19)

Roth places Mayr's views on homology as firmly within the phylogenetic framework, which makes sense of his comments on bat and bird wings. While he may have regarded the forelimb structures as homologous, their change into wings was convergence. Roth is implying that different developmental trajectories might determine the convergence in wing morphology, and she takes a fairly liberal view on what genetic causes might be at the root of such differences.

Central to Kirschner and Gerhart's (2010) position is the idea that the huge conservation across taxa in genes, which yields molecular homologies, put constraints on selection in terms of creating new forms. This can readily be reinterpreted as a statement that molecular homologies constrain available ontogenetic options, which in turn limits the choices for selection. There is no need to see selection as directly creating form. Their view also allows the claim that evolution has not been uniform in tempo and mode over time, and there have been periods of stasis. A major innovation was the prokaryotes *inventing* DNA, RNA, and protein synthesis around 3 billion years ago, followed 2 billion years ago by the eukaryotes *inventing* the compartmentalized cell. These things are basically unchanged today and lead to a crucial point that conserved "core processes represent the basic machinery of the multicellular organism, specified for its specific diversified functions by regulatory control" (Kirschner and Gerhart 2010: 260). These core processes can be organized hierarchically to deliver development—understood as differentiation over time—and those hierarchies can turn into composite processes that are conserved within lineages. As an example, they reference the segmentation system in insects that is maintained within the entire arthropod lineage. And while a few "gene products may be unique to a trait, such as the specific forms of keratin in hair and skin, or ion channels in the heart, most aspects of a trait are produced by some combination of the shared core processes" (Kirschner and Gerhart 2010: 261).

Thus, Kirschner and Gerhart's idea is that there are core developmental modules that can be differently regulated to produce phenotypic variation and only require a minimal extra genetic variation to continue into novel specialization. These core modules are subject to mutation, but most mutations would be deleterious as they would disrupt this essential platform. Here Kirschner and Gerhart see an evolutionary efficiency in that new builds reliant on entirely new genetic infrastructure with regard to variation would be subject to much more destabilizing mutation, or lethality, and this would considerably slow evolutionary response.

If we apply this kind of thinking to Roth's discussion, we can see that the forelimb homology is likely underpinned by the same developmental architecture and that the key differences between bat and birdwings are delivered by differences in regulatory genetics that are specific to bat and bird lineages.

6.3.2 Kirschner and Gerhart's Theory of Facilitated Variation

Kirschner and Gerhart link their developmental module model to *evolvability* which they see as consisting of a selection and a variation component. The former is a standard view of the phenotype fitting to contingent environmental problems. The latter they see as a capacity to produce phenotypic variation in response to genotypic variation. They see the ability to maximize phenotypic variation relative to genotypic variation while minimizing lethality and producing phenotypic variation relevant to previously unencountered conditions at the lineage level as central to evolvability. The variation component is underpinned by:

1. Exploratory processes.
2. Weak linkage.
3. Compartmentation.

Exploratory processes refer to the production of phenotypic variation based on a small amount of genetic variation and then selection at a molecular and cellular level. Kirschner and Gerhart give the example of microtubule formation:

> Preprogrammed variation followed by selection is widespread in cellular systems. Dynamic assembly and local stabilization of microtubules occurs in virtually all cells of every multicellular organism, as well as in unicellular eukaryotes. In this case an input of energy makes each microtubule polymer unstable in a very special way. Microtubules undergo excursions of growth and shrinkage in a seemingly pointless exploration of cytoplasmic space. Yet because of the fast turnover of structure, local stabilization at the end of a microtubule can appear to re-direct the entire array so that it is polarized toward the stabilizing signal. The localized stabilization agents select a particular microtubule sub-population in a certain region of the cell. The resulting polarized array of microtubules is not generated directly, but selected one microtubule at a time from a rapidly changing unpolarized array. (Kirschner and Gerhart 2010: 263)

They compare this process to that of adaptive immunity. A key point for them is that this machinery allows a huge amount of diversity, much of which is discarded in current specific contexts. But that diversity is potential for evolution. By this, what they actually mean is that more than one phenotypic end is potentially the case in this set-up. They do not mean it is unconstrained and in fact mean precisely that for this diversity to find use a new constraint or system of constraints, would have to come into being in order to effect novel stabilization. This will likely be due to minimal but novel changes in genetic variation that make conserved developmental packages deliver novel traits, as above.

Later they discuss other exploratory systems, including neuronal systems that, if they do not find muscle tissue simply die off, in much the same way as unstabilized microtubules. This kind of architecture is entirely conditional. There are clear energy budgets attached to such exploratory growth and clear signals to continue development.

What of *weak linkage*?

> Weak linkage refers to specific biochemical features of information systems in biology, where signals of low information content evoke complex, preprogrammed responses from the core process. The specifics of the signal have a weak relationship to the specifics of the outcome. The term… also alludes to the typically low energy and low specificity of the interactions between signals and responders. (Kirschner and Gerhart 2010: 265)

What Kirschner and Gerhart are referencing are systems that equivocate between at least two stable states and can be tipped into one or other with minimal input. They directly refer to these systems as switch-like, and again this all fits neatly with conditional architecture descriptions and more formal accounts of information theory advocated in this book (Sect. 5.2). Such systems can be linked together to enable great flexibility and responsiveness, as we see in brains. This is to be contrasted with strong linkage where a datum input would have a unique relationship with a context. Their take is that much of biology is hierarchically organized in terms of weak linkage due, in part, to the difficulty of establishing strong linkage and the consequent difficulties of creating multiple, interacting, and strongly linked systems.

Compartmentation refers to a form of modularity that makes much use of weak linkage. For example, each "spatial compartment in an embryo is defined by a small set of unique selector genes, which encode transcription factors or signalling molecules that are expressed uniquely in that compartment. The selector gene can then 'select' any other gene to be expressed or repressed in its compartment" (Kirschner and Gerhart 2010: 267). This allows differentiation, modulation of timing and amplitude of response etc. and all in parallel.

Kirschner and Gerhart discuss the three inventions of wings in vertebrates, within the pterosaurs, birds, and bats, noting they each involved a different compartmental modification to achieve a similar morphology that responded to aerodynamic requirements. In doing this, they come full circle to the comments by Mayr on homology. Under the model proposed by Kirschner and Gerhart these wings rely upon conserved processes coming under different regulations within a relevant compartment during development. While there is much conservation here, this does not quite do away with Mayr's original observation that this is due to convergent evolution. Indeed, this is a case of convergent evolution as a result of selection upon three different occasions for changes in the control of conserved developmental processes. What is different is the extent of the claim. Mayr was possibly thinking about genes all the way up and imagining that would be hard to engineer. Kirschner and Gerhart agree that would be hard, and they are suggesting a core homology of basic developmental structure that needed only a little change in control to give it direction. Those changes converged on the same solution. Undoubtedly Kirschner and Gerhart would say the developmental system they hypothesize is massively constrained and makes wings more likely, but Mayr could well have imagined constraints in his original version of events.

This last comment is not an attempt to save the credibility of Mayr but rather to point to the ease with which some theorists claim large shifts away from standard theory and historically standard accounts. The developmental biology cited here is wonderful and exciting and certainly unknown in the late synthesis period when Mayr commented on wing convergence. But it is not in any sense challenging the framework of the Modern Synthesis—indeed it relies upon the notion of selection, genes, and information, all of which are common currency. But more than that the complexity of the developmental process was undoubtedly bracketed but in a way that allowed its later completion. It was assumed not to falsify the central tenets of the Modern Synthesis, and the theory of facilitated variation indeed does not falsify the synthesis. This does not make this position in developmental biology less or more interesting; it is merely consilient.

6.3.3 Modular Development and Gene Centrism

Newman locates cell differentiation as a unicellular basis for the onset of multicellular developmental processes (Newman 2010).

> The biosynthetic state of all cells are determined by the dynamics of transcription factor-mediated gene regulatory networks (GRNs)... Such networks, containing feedback and feed-forward loops by which the transcription factors promote and suppress their own and each other's synthesis, exhibit multistability... The systems can thus switch among discrete states, the number of states always being much smaller than the total number of genes in the organism's genome. Since the genes that specify nontranscription factor proteins and regulatory RNAs are themselves subject to transcriptional control, the alternative stable states of the GRNs specify cell types distinguished by extensive biosynthetic differences. (Newman 2010: 281–2)

Newman makes clear that complex regulatory networks are going to produce multistable behavior as a matter of course and that such stability is not to be regarded as adapted. Newman sees evolutionary processes as reducing this propensity to flip between states by suppressing costly forms that might disrupt the overall organism to which the cell types are contributing.

Newman claims that the formation of phenotypes also relies upon dynamic patterning modules, which allow the emergence of developmental patterns (spatial and temporal sequencing through ontogeny) and the ordering of cell types produced by complex regulatory networks (Newman 2010, 2017). Dynamic patterning modules are "associations of specific gene products" and the "physical effects they are capable of mobilizing in the context of cell aggregates" (2017: 192). This leads to the following view:

> In contrast to the idealistic view implicit in the (Modern Synthesis) that genes must collectively constitute a program for the generation of morphological phenotypes, the (dynamic patterning module) framework... emphasizes the connection of phenotype to the broader material world. Genes act not as conveyors of "information" but as direct and indirect specifiers of building materials, the organizational properties of which inescapably conform to the physical laws appropriate to their composition and scale. What is passed from one generation to the next are not only genes and their variants, but the material systems, including the specific capacities of their gene products to mobilize physical forces and effects in cell aggregates, and the conditional... outcomes of those forces and effects. (Newman 2017: 194–195, parentheses added).

There is much in common here with Kirschner and Gerhart's views on modular development, and indeed this is a recurrent theme in evo-devo (Moczek 2019) related closely to discussions of canalization (Meiklejohn and Hartl 2002; Moczek 2007; Waddington 1942). Canalization is a description of the consistent development of particular traits in spite of environmental and genetic variation, and as such, it is seen as a robustness solution (Chap. 5) such that selection "for developmental processes that successfully produce their target phenotype will therefore implicitly select for canalization" (Meiklejohn and Hartl 2002: 471). What contemporary evo-devo is revealing is the highly conserved detail of developmental modules and the role of modules in the emergence of novel phenotypes as a result of small regulatory changes, ultimately underpinned by genetic variation.

Newman clearly thinks that these kinds of the framework are direct challenges to the Modern Synthesis. While no one is arguing that genes have no role to play, the pervasive interpretation of the Modern Synthesis as gene-centric, which sees genes as total, deterministic causes, is in play here. My claim has been that this is a misinterpretation of the heuristic, idealized use of selfish-gene perspectives, which deliver particular modelling goods, something which some developmentally minded theorists are beginning to recognize (Uller et al. 2020). But, while Uller and colleagues acknowledge the role of idealizations in meeting explanatory tasks demands, they make a case for incorporating developmental arguments into evolutionary ones, thus a case for a new kind of idealization. In part this is because they feel that traditional approaches, of the sort this book is advocating, privilege selection-based accounts to explain the adapted match between an organism and the environment. Sometimes that fit maybe delivered by, for example, learning processes.

It is undoubtedly true that fit might be delivered by learning or other plastic and responsive processes. It is true because it is almost a truism. *Adaptive* simply refers to that kind of phenomenon. Here we see the elision of Spencer and Darwin (Sect. 6.2.1), leading Uller and colleagues to assume that the business of accounting for evolution is the same as the business of accounting for development. And this conflation, while historically grounded as Depew has discussed (Depew 2017), is also a consequence of a laser-like focus upon the creation of form in the sense of its construction within an individual. Uller and colleagues are quite clear that they see a role for transformational views in evolutionary theory, but in so doing, they are aligning themselves with pre-Darwinian instincts about change across time (Chap. 4).

Uller and colleagues do suggest that an evo-devo perspective, which they see as a blending of transformational and selective accounts, is essential for explaining evolutionary novelty and in particular the emergence of new kinds of individuals. They suggest evo-devo as a new kind of idealization for this task because the selection of new kinds of individuals incorporates the selection of new developmental processes, and they assume an interaction between selection and developmental processes. But as we have seen, much of evo-devo relies upon highly conserved developmental modules, which constrain dynamic physical processes that achieve certain forms of equilibrium and which can be modified by the selection of novel genetic regulation. If the explanandum is phenotypic novelty, then, by this evo-devo account, changes to gene regulation are the necessary first step and thus privileged in that minimal sense. If the explanandum is the existence of developmental modules, they too are a product of highly conserved genes that enable the physical processes required for development to occur, in keeping with, for example, Dawkins' catalyst argument (Sect. 3.4). Here again, genetics will be privileged in the most minimal of senses. What privilege does not mean is that one component of the account is regarded as more or less important, which is a natural error of colloquial inference to be drawn here. Nor is this a stronger claim about necessary *and* sufficient conditions. Instead, it is simply a claim that the existence of DNA code is a necessary condition in the development of a phenotype. Furthermore, the emergence of genetic variation is a necessary condition for the evolution of populations with novel developmental trajectories.

Newman has made all this clear with his account of genes as specifiers of building materials that have their own intrinsic properties that lead to organizational effects. But he tells us this means that genes are not conveying information. Technically this is right, as *information = data + context* (Sect. 5.2) and I am very clearly arguing that genes operate within a context. Genes, as functional portions of DNA code, are data that find their context in the presence of RNA and the rest of the protein synthesis machinery, and in the context of other genes and so forth. They are functional portions of DNA because they change the state of the system, or context into which they are inserted, and that leads to a variety of downstream effects that are packaged as development. Newman's term, *specifier*, is itself perhaps inspired by its use in linguistics to denote determiners (e.g., a, the, this), quantifiers (e.g., some, every) and possessives (e.g., Darwin's, her's). Thus, genes determine particular meanings within specific contexts by leading to particular routes of interpretation, reducing ambiguity or uncertainty. This is the very core of information theory.

Genes may also be seen as privileged due to their role in inheritance. This would explain Newman's comment above, noting that more than genes are passed from one generation to the next. Genes are of interest to Modern Synthetic accounts because of their high-fidelity transmission across generations. As crucial inputs into developmental contexts, this property is a key part of ensuring stable developmental outcomes within lineages. As fidelity relates to consistency, privilege again denotes a necessary but not sufficient condition when packaged with inheritance.

The view I am espousing is, of course, in line with Mayr's statements about DNA codes and programs (Chap. 4). Where Mayr is thought to advocate a developmental program written into the genes, he is in fact noting the role of genes as data in a process. The modular approach to development from evo-devo tells us that development unfolds in a sequence that is organized in modular packages that can be rearranged to deliver new outcomes. This kind of organization is found in many computational programming languages and coding approaches. For example, the open-source statistical project R delivers flexibility because numerous individuals have created packets of code that deliver subtasks that can be rearranged to deliver various different kinds of analysis (R Development Core Team 2009). Evo-devo is highly programmatic in its outlook, but it is also sophisticated in that it provides proximate detail of the mechanistic dynamics at the cellular level and understands mechanisms as constraints on energy that achieve work (Bechtel and Bich 2021).

It would appear that the decision to bracket development during the formation of the Modern Synthesis—if it was a decision—but to broadly characterize it in the way that Waddington, Mayr, and others did provided just the right theoretical constraints in which evo-devo could thrive. Those brackets are open research questions (Griffiths and Gray 2000).

6.4 What Is Left?

Adherents to the Modern Synthesis think of development as the process of growth, differentiation, and specialization in an organism. It begins, in a sexually reproducing species, at conception and continues, building a reproductively viable adult. Evolution, on the other hand, is the process of population change. A population of organisms will change in terms of the relative abundance of particular kinds of organisms, where kinds are determined by traits.

Without committing to a level or unit of selection, we can see that these two processes are different. Moreover, we can see that natural selection might test the viability of an organism in a particular setting at any stage in its life, from conception onward. In this way, selection could shape the available developmental profiles, the available trajectories, and we should expect this to lead to a considerable degree of canalization to maintain robustness (Meiklejohn and Hartl 2002).

This view of development does not commit us to a particular view on how it works, but for it to be open to selection, then development must be heritable to some useful extent. We should therefore expect some high fidelity transmission of developmental capacity or tendency. The modular approaches discussed in Sect. 6.3 meet this requirement because modules are dependent upon highly conserved genes, and their organization into different traits between lineages is due to differences in regulatory genes. The physical processes that the developmental modules permit are subject to their own lawlike regularities. We might also expect data sensitivity in keeping with discussions about robustness and plasticity (Chap. 5). In other words, we might anticipate that development, where plastic, responds to data that predicts future circumstances well, but that this responsiveness is not limitless but usefully parameterized.

On this last point, we might initially think that development is temporally protracted because it simply takes a while to organize resources and construction. There is a sequence to be followed, and sequences of events are the basis of temporal ontology. But the fact that developmental sequences permits decision-making across time facilitates the introduction of plasticity—within limits—and we should perhaps expect degrees of developmental flexibility in the face of environmental perturbation. Each sequential transition is an opportunity for a change in trajectory. Many changes of the trajectory may fail to survive selection, but those that do will introduce data-led pathways or reaction norms to the developmental profile.

When we stand back and look at the array of possible trajectories and their final phenotypic resting points, we see a form of variety; we see variation about a statistically determined mean. These phenotypes are, according to Mayr and many others, where selection acts. Thus, over evolutionary time we should expect selection to explore alternative trajectory space and to render decisions about which pathway is pursued sensitive to data about local conditions, just so long as they predict future conditions. The underlying sensitivity to data, as well as the paths those data determine, must be heritable for evolution to work, and in a mechanistic world view, that means a physical cause that underlies the phenotypic phenomena we wish to explain. The phenotype cannot emerge *de novo* from nothing. Moreover, over evolutionary time, for selection to winnow the possible trajectory space to a small sub-set of reaction norms that fit the organism to its contingent realities, the causes of the phenotype must be heritable.

The heritable, underlying physical causes of reaction norms will themselves be subject to change, and those changes can introduce more or new possibilities into the developmental trajectory space.

While all this might seem reasonable in light of the preceding discussion, reaction norms have not escaped the scrutiny of evo-devo scholars and those advocating for an extended synthesis.

6.4.1 Reaction Norms

In the early 1900s, during the debate between the biometricians and Mendelians, Johannsen made the claim that there was no inheritance of individual variation between generations and in so doing, he challenged the pervasive transmission view of inheritance (Sect. 2.5.1, Nicoglou 2018). He replaced the traditional view with the genotype view in which what was passed onto the next generation was something that could determine characteristics, but not the characteristics themselves. Indeed, Johannsen introduced a number of key terms for biology during this period (Johannsen 1911):

> I have proposed the terms "gene" and "genotype" and some further terms, as "phenotype" and "biotype," to be used in the science of genetics. The "gene" is nothing but a very applicable little word, easily combined with others, and hence it may be useful as an expression for the "unit-factors," "elements" or "allelomorphs" in the gametes, demonstrated by modern Mendelian researches. A "genotype" is the sum total of all the genes in a gamete or in a zygote. (Johannsen 1911: 132–133)

> All "types" of organisms, distinguishable by direct inspection or only by finer methods of measuring or description, may be characterized as "phenotypes." (Johannsen 1911: 134).

Johannsen, while aligning genes with the Mendelians and particulate theories, did not have a theory about the material basis of genetics at this point, and he was fastidious in his use of his new concepts, preferring to refer to genotypical differences as it was not possible to truly know the genotype (Johannsen 1911; Nicoglou 2018). In this way, Johannsen was creating an idealization to make sense of his and other scientific findings (Potochnik 2020). He based this idealization in particular on Weismann's separation of the germline from the soma and also Woltereck's concept of the *Reaktionsnorm*, more commonly referred to as either the reaction norm or norm of reaction in English speaking communities.

Woltereck had noted that individual Daphnia can vary phenotypically within a clonal line due to different environmental inputs and that the nature of that phenotypic variation can differ between non-clonal individuals within the same species. All of the collected phenotypic curves from all of these situations, induced by the relevant environmental inputs, capture the norm of reaction for that species (Nicoglou 2018; Stearns 1989). The reaction norm is inherited, and Johannsen (1911) saw the genotype concept as compatible with the reaction norm in that the genotype determined the reaction norm and enabled its inheritance. Nicoglou (2018) points out that Johannsen's view differs from Woltereck's in that it was more genetically deterministic, whereas Woltereck understood the phenotype to be constructed in part by the environment. Nicoglou goes onto argue that Johannsen's position, combined with his genotype-phenotype distinction paved the way for a much harder view, following Gould (2002), in which nature and nurture were duly seen as separable contributors to form.

Stearns (1989) briefly relays the history of reaction norms, noting it was not widely accepted as a part of biology for some time. Stearns tells us that Wright saw this kind of plasticity as uncoupling the phenotype from the genotype to enable adaptive fit without the need for new genetic variation, while Schmalhausen regarded the norm of reaction as historically stabilized plasticity that enabled some fit with different environments. Dobzhanksy argued that it was the norm of reaction of the organism to the environment that changed during evolution, in other words, it was the plastic phenotype that was the focus of selection, and this brought gene frequency changes. In doing this, Dobzhansky brought the reaction norm to the population level such that the study of the adaptive norm was the study of all the genotypes, rather than of their phenotypes (Sarkar 1999). Dobzhansky and Spassky claimed that the adaptive norm represented approximately 95% of the population (articulated as standard deviations about the mean fitness), while the genetic load (or burden) were the approximately 2% at the extreme left of the x-axis and the genetic elite the approximately 2% at the extreme right. This adaptive norm was understood as a generalist plastic strategy.

> Natural populations of Drosophila, man, and presumably of all sexual, diploid, and outbreeding organisms contain a multitude of genotypes. A majority of these genotypes make their carriers tolerably well adapted to survive and to reproduce in the environments which the population frequently encounters in its natural habitats. The array of such genotypes constitutes the adaptive norm of the species or population. Some genotypes yield, however, low fitness in the habitual environments; these compose the genetic load of the population. And finally, some genotypes confer a fitness distinctly above the mean of the adaptive norm; these are the genetic elite of the population. The ill-adapted genotypes which arise in the population when it follows its normal breeding system are the expressed genetic load; deviations from the normal system toward greater in-breeding bring to the surface a part of the normally concealed genetic load consisting of recessive genes and gene combinations. Populations that are normally outbred have concealed loads which are much greater than the expressed loads. (Dobzhansky and Spassky 1963: 1467)

Recently Sultan has inspected the concept of the reaction norm from the perspective of evo-devo (Sultan 2019). She opens with the following statement:

> Norm of reaction studies carried out in diverse systems and conditions have provided two key insights. First, genotype and environment co-determine individual phenotypes and hence, at the population level, the adaptive variation on which natural selection acts. Second, nonparallel reaction norms cause the size and rank order of genotypic differences to vary among environments, making the genetic consequences of selection environment dependent. (Sultan 2019: 109)

She is keen to understand development as an interaction between genotype and environment and to contrast this with what she regards as the traditional developmental program approach (Sect. 6.3.3) and she follows Newman's view that the Modern Synthesis saw genes as conveying developmental information (Newman 2010, 2017).

Sultan takes aim at what she describes as the fitting of the environmental context-specificity of phenotypic expression to a genetic or developmental program view. "Evolutionary biologists have reconciled their awareness of genotype-environment interactions with a genetically deterministic model for development by defining the norm of reaction itself as a self-contained 'property of the genotype'" (2019: 110). Clearly, this view has its antecedent in Johannsen's articulations of the relationship between genotype and reaction norm and Sultan, in common with Nicoglou (2018), regards this as yielding control to genes while permitting context dependent phenotypic outcomes. But Sultan does not make this historical link to Johannsen but instead claims that it was the mid-twentieth century period that saw the emergence of the genetic control view. She discusses a number of views from this period that appear to see the array of phenotypic expression as under direct genetic control, such that there are plasticity genes. Since the 1990s, this view has changed, she tell us, and gene expression is understood to be context dependent, and environmental factors have a role in regulatory complexes. Nonetheless, "sophisticated approaches to norm of reaction parameters… continue to presuppose that the reaction norm is a fixed genotypic property" (Sultan 2019: 112). She follows this with a quote from a glossary by Chevin and colleagues in which they define a reaction norm as "the expected phenotype of a given genotype as a function of the environment" (Chevin et al. 2010: 2).

It is worth inspecting what Chevin et al. mean by function in the preceding quote. Here is a fuller expression from the same paper:

> *Biological impact of the environment.* This links the environment to the biology of individuals in the species. Phenotypic plasticity describes the direct impact of the environment on the development of individual phenotypes. It may involve morphological, physiological or behavioural responses, which can occur on different time scales. For continuous environmental variables, plasticity usually is modelled using reaction norms, where the phenotype of a given genotype is plotted as a function of the environment. We focus on linear reaction norms for simplicity, although reaction norms can be non-linear... With linear reaction norms, the slope b quantifies the degree of plasticity. The environmental sensitivity of selection, B, measures how the optimum phenotype changes with the environment, which for simplicity we also assume is a linear relationship. With no cost of plasticity, populations with b closer to B are likely to persist under higher rates of environmental change. (Chevin et al. 2010: 3)

A linear relationship is being assumed for simplicity such that the phenotype (Y) is a result of $f(x)$, where x is the environmental exposure of the genotype, e.g., a nutrient gradient, under some function (f), such that $Y = f(x)$. The functional parameters will determine the slope (the measure of plasticity) and intercept (the response value in the average environment), and there will of course be residual error about the fitted line (Dingemanse et al. 2010). While regression models like this are causal to some extent, they do not denote the causal mechanism, but merely the crucial parameters for deriving a prediction from x to Y. Sultan's objection is that the genotype is assumed to account for the entirety of the function (f) and seems to have causal primacy and control in this equation. But Chevin and colleagues in fact suggest morphological, physiological, and behavioral processes as possible causal factors at this point. They presumably do this because the paradigm rests upon an assumption that the genotype is constant across all conditions of x. It is true that Chevin and colleagues are trying to understand how the limits of plasticity under environmental change might affect the future survival of particular species, and they recruit standard views on fitness to begin that project, but at no point are they arguing for purely genetic control of plasticity.

Put another way, creating models to try and understand how plasticity may enable the persistence of a genotype in changeable environments does not impose the view that genes have total, deterministic control of the expressed phenotype. It is rather the opposite—the developmental and physiological systems that have been selected for afford extra informational capacity for the organism beyond that yielded in the genotype. Where there is gene centrism it is in the perspective that this plasticity preserves the genotype, maintaining high fidelity transmission across generations. Plastic mechanisms are built by genes, but those mechanisms are designed to produce their own bespoke responses within parameters. Here there is a difference between *coding for the mechanism* and *coding for the work done*. The Modern Synthesis made no statement in support of the latter view. Sultan, and many others, conflate the two when interpreting the Modern Synthesis.

Sultan offers an alternative to the programmatic view she believes is in play:

> Empirically based insights regarding the formative role of complex genotype-environment interaction suggests a re-evaluation of genetic control as a causal principle. Developmental outcomes are shaped by multiple types of information and not by DNA sequence alone... (P)henotypes emerge from the real-time regulatory interactions of the evolved genotype with transient environmental and/or epigenetic influences that occur at timescales from within a generation to several or many generations. These factors cannot be pulled apart as individual causes, because they contribute interactively to phenotypic expression and hence to norms of reaction. Consequently, phenotypic outcomes can be seen to arise from 'distributed control' in each generation. (Sultan 2019: 117)

This view leads Sultan to suggest that experimental work on phenotypic expression must test specific genotypes in more than one environmental context, which requires an understanding of relevant contexts not dissimilar to the work of Woltereck in the 1900s. She believes such designs will reveal strong interaction effects between genotypes and "inherited environmental and epigenetic factors" (2019, 118) that have a major influence on the expression of phenotypes, and she refers to work on collective behavior in collectives with no central control (Gordon 2016).

Gordon discusses the nature of algorithms that regulate collective behavior in multiple systems. She is clear that the use of the term algorithm is appropriate but should be understood in the absence of a programmer (she is no modern-day Paley), and she lays the source of design in natural systems within evolutionary processes. Sultan cites Gordon stating that "in a distributed process, the role or behaviour of an individual or component is determined by its interactions with others rather than its inherent attributes." Below I quote the full passage this is drawn from:

> Despite this diversity, the algorithms used to regulate collective behavior do have some characteristics in common. They tend to be distributed processes. In a distributed process, the role or behavior of an individual or component is determined by its interactions with others rather than its inherent attributes. Some important advantages of distributed processes, flexibility and resilience to disturbance, arise because different participants in the collective outcome are able to respond to changing conditions. For example, task allocation in an ant colony, the distributed process that determines which ant performs which task at a given time, uses an algorithm based on the rate of interaction between ants. An individual ant assesses interactions by detecting the cuticular hydrocarbon profile of each ant it meets. An ant's cuticular hydrocarbon profile provides information about its task because the conditions in which an ant works change the chemistry of the hydrocarbons on its body surface. For example, when harvester ant foragers are out in the sun, the proportion of n-alkanes in their hydrocarbon profiles increases, leading a forager to smell recognizably different from an ant that works inside the nest. An ant uses its recent experience of interactions with other ants in its decisions about what task to perform and whether to perform it actively. Collectively, this allows the colony to regulate the numbers of ants currently engaged in each task in response to the availability and flow of resources and external conditions. For example, when extra food is available, ants engaged in other tasks

> switch tasks to foraging as a result of increased interactions with other ants that found food. (Gordon 2016: 516)

Gordon's example of ants switching between task types is clearly based on the intrinsic properties of individual ants, notably their ability to detect the cuticular hydrocarbon profiles of others. These profiles change due to chemical processes after environmental exposure. Why then does Gordon discount inherent attributes? The simple answer is that she is trying to explain the collective outcome at the ant colony level and is doing that purely via interaction-based explanations. In her algorithmic idealization, she is prepared to discount intrinsic causes of conditional decision switches in individual ants to focus on the outcome of multiple such switches in the presence of environmental exposures. She has a particular task demand that this suits. But she is explicitly telling us that the intrinsic ability of ants to detect these chemicals is a necessary condition to implement this algorithm in this case. Remember, Gordon's ambition in this passage is to explore explanatory similarities across collective action accounts.

I am sure that Sultan would similarly lay claim to this generality and idealization strategy in keeping with standard understandings of generating causal models (de Regt 2017; Potochnik 2020). Her commitment to complex controlled experimental design suggests she is pursuing an epistemological strategy here. But it all becomes a little rhetorical because her initial complaint is about a reliance upon the idea of a genetic or developmental program (she uses both terms interchangeably), and her solution is to specifically invoke an algorithmic view of development. Implemented algorithms are programs. She does not see that the talk of programs during the formation of the Modern Synthesis was suitably broad to simply allow her position and that the bracketing of development was not unduly restrictive.

Sultan's chapter title references the unscripted reaction norm, and this is perhaps the biggest clue to her discontent. It is possible that she believes the DNA program idea is utterly deterministic in a Laplacean sense. And yet, as we have seen (Sect. 3.4) even full-throated gene-level theorists such as Dawkins committed to a catalyst view of genes in development. So, even if one were to directly argue for genes for plasticity, those genes are only conceptualized as a part of a long-chain development process, without context they are inert, without context they have no informational value. Gordon's arguments can be reinterpreted in this information theoretic fashion, with her collective ants coordinated at the colony level by relevant data inputted into relevant subsystems of the colony, in this case, individual ants. The chemical changes due to environmental exposure are inputs into other ants, who in turn produce relevant behaviors which input into the colony. This high level of cooperation or coordination means that the colony can be seen as an individual (Queller and Strassmann 2009) and as Gordon herself notes, the algorithm that captures this coordination has stabilized through evolutionary transitions.

6.5 Summary

In this chapter I have made the following key claims. First, while developmental biology may have been black boxed during the Modern Synthesis the parameters of that black box were sufficiently generous to readily incorporate the developmental insights from evo-devo discussed above. Second, the role of the gene in development has always been one of a necessary but not sufficient condition for Modern Synthesis. Third, and related to the second, the tendency to package gene centrism in colloquial information talk has led to a view of genetic determinism due to a notion of complete instruction or representation at the gene-level, something that is not warranted. The application of a data led view such that information is the functional relationship between data and context immediately clears this misconception away and again makes room for evo-devo accounts. I have also pointed out that the detail of the evo-devo accounts presented here can be clarified by taking a precise view on data and its relation to context, and this helps to understand the nature of idealizations that are in use for particular theorists.

This chapter has in no way surveyed the whole of evo-devo. My intention has been to focus on some key examples of fascinating work in this area and work that has been recruited to challenge the Modern Synthesis. Given this work has been held up in support of extension, my claim is that the analysis I have subjected it to here will generalize to other examples used in the same cause. Nonetheless, in the following chapters, I turn my attention to epigenetics and niche construction theory, both of which have contributed to the developmental challenge to the Modern Synthesis.

References

Bechtel W, Bich L (2021) Grounding cognition: heterarchical control mechanisms in biology. Philos Trans R Soc B Biol Sci 376(1820):20190751. https://doi.org/10.1098/rstb.2019.0751

Chevin LM, Lande R, Mace GM (2010) Adaptation, plasticity, and extinction in a changing environment: Towards a predictive theory. PLoS Biol 8(4). https://doi.org/10.1371/journal.pbio.1000357

Dawkins R (1976) The selfish gene. Oxford University Press, Oxford

de Regt HW (2017) Understanding scientific understanding. Oxford University Press, Oxford

Depew DJ (2017) Natural selection, adaptation, and the recovery of development. In: Huneman P, Walsh DM (eds) Challenging the modern synthesis: adaptation, development, and inheritance. Oxford University Press, pp 37–67

Dingemanse NJ, Kazem AJN, Réale D, Wright J (2010) Behavioural reaction norms: animal personality meets individual plasticity. Trends Ecol Evol 25(2):81–89. https://doi.org/10.1016/j.tree.2009.07.013

Dobzhansky T, Spassky B (1963) Genetics of natural populations. Xxxiv. Adaptive Norm, Genetic Load and Genetics 48(November): 1467–1485

Eronen MI (2015) Levels of organization: a deflationary account. Biol Philos 30(1):39–58. https://doi.org/10.1007/s10539-014-9461-z

Gordon DM (2016) The evolution of the algorithms for collective behavior. Cell Syst 3 (6):514–520. https://doi.org/10.1016/j.cels.2016.10.013

Gould SJ (1977) Ontogeny and phylogeny. The Belknap Press of Harvard University Press

Gould SJ (2002) The structure of evolutionary theory. The Belknap Press of Harvard University Press

Grafen A (1984) Natural selection, kin selection and group selection. In: Krebs JR, Davies NB (eds) Behavioural ecology, 2nd edn. Blackwell Publishers, New York, pp 62–84

Griffiths PE, Gray RD (2000) Darwinism and developmental systems. In: Oyama S, Griffiths PE, Gray RD (eds) Cycles of contingency: developmental systems and evolution. MIT Press, London

Hull DL (1980) Individuality and selection. Annu Rev Ecol Syst 11(1):311–332. https://doi.org/10.1146/annurev.es.11.110180.001523

Johannsen W (1911) The genotype conception of heredity. Am Natural 531(XLV):129–159

Kirschner MW, Gerhart JC (2010) Facilitated variation. In: Pigliucci M, Müller GB (eds) Evolution: the extended synthesis. MIT Press, London, pp 253–280

Laland KN, Sterelny K, Odling-Smee J, Hoppitt W, Uller T (2011) Cause and effect in biology revisited: is Mayr's proximate-ultimate dichotomy still useful? Science (New York, NY) 334 (6062):1512–1516. https://doi.org/10.1126/science.1210879

Laland KN, Uller T, Feldman MW, Sterelny K, Mu GB, Uller T, Moczek A (2015) The extended evolutionary synthesis: its structure, assumptions and predictions. Proc R Soc B Biol Sci 282. https://doi.org/10.1098/rspb.2015.1019

Love AC (2017) Evo-devo and the structure(s) of evolutionary theory. In: Huneman P, Walsh DM (eds) Challenging the modern synthesis: adaptation, development, and inheritance. Oxford University Press, Oxford, pp 159–187

Mayr E (1961) Cause and effect in biology. Science 134:1501–1506. https://www.ifor.math.ethz.ch/education/bachelor/lectures/fs2011/math/Biomath/Mayr.pdf

Mayr E (1984) The triumph of evolutionary synthesis. Times Literary Supplement, November(2), 1261–1262

Mayr E (1992) Controversies in retrospect. In: Futuyma DJ, Antonovics J (eds) Oxford surveys in evolutionary biology: Volume 8. Oxford University Press, Oxford, pp 1–35

Mayr E (2002) What evolution is. Phoenix

Meiklejohn CD, Hartl DL (2002) A single mode of canalization. Trends Ecol Evol 17(10):468–473. https://doi.org/10.1016/S0169-5347(02)02596-X

Moczek AP (2007) Developmental capacitance, genetic accommodation, and adaptive evolution. Evol Dev 9(3):299–305. https://doi.org/10.1111/j.1525-142X.2007.00162.x

Moczek AP (2019) The shape of things to come: evo devo perspectives on causes and consequences in evolution. In: Uller T, Laland KN (eds) Evolutionary causation: biological and philosophical reflections. MIT Press, London, pp 62–80

Müller GB (2007) Six memos for evo-devo. In M. D. Laubichler & J. Maienschein (Eds.), From embryology to evo-devo: a history of developmental evolution (pp. 499–524). https://doi.org/10.7551/mitpress/3128.003.0018

Newman SA (2010) Dynamical patterning modules. In: Pigliucci M, Müller GB (eds) Evolution: the extended synthesis. MIT Press, London, pp 281–306

Newman SA (2017) Toward a nonidealist evolutionary synthesis. In: Huneman P, Walsh DM (eds) Challenging the modern synthesis: adaptation, development, and inheritance. Oxford University Press, Oxford

Nicoglou A (2018) The concept of plasticity in the history of the nature-nurture debate in the early twentieth century. In: Meloni M, Cromby J, Fitzgerald D, Lloyd S (eds) The Palgrave handbook of biology and society. Palgrave Macmillan, London, pp 97–122. https://doi.org/10.1057/978-1-137-52879-7_5

Noble D (2011) Neo-Darwinism, the modern synthesis and selfish genes: are they of use in physiology? J Physiol 589(Pt 5):1007–1015. https://doi.org/10.1113/jphysiol.2010.201384

Otsuka J (2019) Ontology, causality, and methodology of evolutionary research programs. In: Uller T, Laland KN (eds) Evolutionary causation: biological and philosophical reflections. MIT Press, London, pp 247–264

Pigliucci M (2007) Do we need an extended evolutionary synthesis? Evol Int J Org Evol 61 (12):2743–2749. https://doi.org/10.1111/j.1558-5646.2007.00246.x
Potochnik A (2020) Idealization and many aims. Philos Sci 87(5):933–943. https://doi.org/10.1086/710622
Queller DC, Strassmann JE (2009) Beyond society: the evolution of organismality. Philos Trans R Soc B Biol Sci 364(1533):3143–3155. https://doi.org/10.1098/rstb.2009.0095
R Development Core Team (2009) R: a language and environment for statistical computing. http://www.r-project.org
Roth VL (1984) On homology. Biol J Linn Soc 22:13–29
Sarkar S (1999) From the Reaktionsnorm to the Adaptive Norm: The Norm of Reaction, 1909–1960. Biol Philos 14:235–252
Scholl R, Pigliucci M (2015) The proximate–ultimate distinction and evolutionary developmental biology: causal irrelevance versus explanatory abstraction. Biol Philos 30(5):653–670. https://doi.org/10.1007/s10539-014-9427-1
Sober E (1984) The nature of selection: evolutionary theory in philosophical focus. University of Chicago Press
Stearns SC (1989) The evolutionary significance of phenotypic plasticity. Bioscience 436 (August):1–10
Sultan SE (2019) Genotype-environment interation and the unscripted reaction norm. In: Uller T, Laland KN (eds) Evolutionary causation: biological and philosophical reflections. MIT Press, London, pp 109–126
Uller T, Feiner N, Radersma R, Jackson ISC, Rago A (2020) Developmental plasticity and evolutionary explanations. Evol Dev 22(1–2):47–55. https://doi.org/10.1111/ede.12314
Waddington C (1942) Canalization of development and the inheritance of acquired characters. Nature 3811(14):563–565. http://chd.ucsd.edu/_files/winter2009/waddington.nature.pdf
Wright S (1931) Evolution in Mendelian populations. Genetics 16(March):97–158. https://doi.org/10.1007/BF02459575

Epigenetics

7

Abstract

The epigenetic processes of gene regulation and the inheritance of epigenetically caused phenotypic effects across limited generations have been recruited by those seeking to challenge the Modern Synthesis. Unsurprisingly it is the extra-genetic nature of these effects that have been the focal concern, suggesting to advocates of the Extended Evolutionary Synthesis a way to avoid gene centrism in developmental and evolutionary explanations, as well as possible use of developmental processes in evolutionary accounts. This last idea has motivated an argument for a more inclusive view of inheritance that includes not only genes and epigenetic processes but also others such as social learning. I discuss all these views in the context of information theory. This leads me to make a statement about the dependencies between genetic and extra-genetic levels of organization that is in keeping with the Modern Synthesis.

Keywords

Central dogma of molecular biology · Soft inheritance · Hard inheritance · Data · Information · Epigenetics · Epigenesis and preformation · Methylation · Avatars of information · Inclusive inheritance · Systemic realization

7.1 Introduction

In the previous chapter, I discussed evolutionary developmental biology, or evo-devo, and in particular work on the ontogenetic emergence of phenotypic novelty and the conception of reaction norms. Both topics have been mobilized by advocates of an extended synthesis and have been used to challenge what is termed gene centrism. A part of my counter to this criticism has been that gene centrism, understood as genetic determinism, is an interpretation of the Modern Synthesis derived because of colloquial and imprecise uses of the term *information*. Once due

T. E. Dickins, *The Modern Synthesis*, Evolutionary Biology – New Perspectives on Its Development 4, https://doi.org/10.1007/978-3-030-86422-4_7

attention is paid to what information is, genes are readily understood as necessary inputs into an overall development context or system. They convey no content, no instruction and no representation, and they merely cause state changes in aspects of the overall development context. In this way, they have an informational role or function.

I have separated the discussion of novelty and reaction norms from other aspects of what Love has termed the developmental challenge to the Modern Synthesis (Love 2017, Chap. 1). In this chapter, I deal with a major developmental challenge, that of epigenetics. Epigenetic processes have clear implications for the development of form but potentially also for an inheritance, as I discuss in this chapter. While inheritance is used in natural selection, much of the focus upon inheritance from this developmental challenge is oriented toward explaining persistent form via extra-genetic processes and, in this way, is perhaps seen as a claim to pluralism. As previously, I interpret the literature from epigenetics and arguments for extension using information theory, as described in Chap. 5. I begin this analysis with a discussion of the central dogma of molecular biology, which forms a foundation for my subsequent comments on epigenetics, epigenetic challenges and then claims for inclusive or extra-genetic inheritance.

7.2 The Central Dogma of Molecular Biology

The precise molecular details of genetics were perhaps irrelevant to the Modern Synthesis during its formation due to its focus upon population genetics, but one key element of the understanding of molecular biology was not, and that was captured as the *central dogma of molecular biology* (Crick 1970). This amounted to a statement that the data contained within genetic material (DNA) flowed only one way, from genotype to phenotype, and no data could be captured at the phenotypic level and passed back to change DNA coding. While phenotypic variation could be acquired throughout life, these acquired characteristics did not affect DNA sequence, and therefore they could not be genetically inherited. This marked the demise of any lingering Lamarckism in evolutionary biology and the Modern Synthesis especially (Gould 2002). The central dogma is the purest statement of the hard inheritance at the core of the synthesis (Mayr 1982).

Crick's classic 1970 *Nature* paper reflected on the central dogma, an idea he had articulated some 12 years previously (Crick 1958, 1970). He begins by outlining the original idea and noting that it was a clear simplification in order to aid understanding created at a time when knowledge of molecular genetics was in its infancy. Thus, the production of a three-dimensional protein after a folding process operating over polypeptide chains had to be bracketed, and, as Crick states, it was simply assumed that "the chain folded itself up" (p.561).[1] This left scientists with a one-dimensional problem of understanding "information transfer from one polymer with a defined alphabet to another" (p. 561).[2] Crick's contribution was to categorize the kinds of transfers that were theoretically possible and to organize them in terms of likelihood. This led to three types of informational transfer.

Type I transfers were:

1. DNA → DNA
2. DNA → RNA
3. RNA → Protein
4. RNA → RNA

The last transfer was hypothesized at the time as RNA viruses were known. More generally, all of these transfers were evidenced directly or indirectly. Type II transfers had no experimental evidence, nor any theoretical demand:

1. RNA → DNA
2. DNA → Protein

At the time, Type II was thought to be either rare or non-existent. Type III was thought to be very unlikely:

1. Protein→Protein
2. Protein→ RNA
3. Protein→DNA

The reason for believing Type III transfers to be unlikely was given by Crick:

> In brief, it was most unlikely for stereochemical reasons, that protein→protein transfer could be done in the simple way that DNA→DNA transfer was envisaged. The transfer protein→RNA (and the analogous protein→DNA) would have required (back) translation, that is, the transfer from one alphabet to a structurally quite different one. It was realized that forward translation involved very complex machinery. Moreover, it seemed unlikely on general grounds that this machinery could easily work backwards. The only reasonable alternative was that the cell had evolved an entirely separate set of complicated machinery for back translation, and of this there was no trace, and no reason to believe it might be needed. (Crick 1970: 562)

[1] Recent developments using deep learning algorithms (recurrent geometrical networks) have enabled rapid and accurate prediction of protein structures from sequence data, giving better insight into how chains fold themselves up (AlQuraishi 2019).

[2] Crick is not using *information* in the formal way discussed in this book. I am using *information* as he does for consistency with his paper in this section. But technically it should be replaced with the term *data*, as in the opening paragraph. I shall return to data and information distinctions later in this chapter.

This led to the statement of the central dogma such that "once (sequential) information has passed into a protein it cannot get out again" (Crick 1970: 562).

Crick emphasizes that the central dogma makes no commitment to transfer and control mechanisms; it only applies to contemporary organisms and that it is a negative statement about what does not occur rather than what does (which is nucleic acids→proteins). Crick's next step is to reorganize the Type I–III transfers according to contemporary knowledge in 1970 as follows:

General transfers

1. DNA → DNA
2. DNA → RNA
3. RNA → Protein

Special transfers

1. RNA → RNA
2. RNA → DNA
3. DNA → Protein

Unknown transfers

1. Protein→Protein
2. Protein→DNA
3. Protein→RNA

General transfers occur in all cells; special transfers do not but can occur in special circumstances, although Crick was doubtful of the third transfer in this set due to limited empirical evidence at the time. Meanwhile, unknown transfers are those that the central dogma hypothesizes will never occur. He makes the case that any discoveries of new cases of special transfer would be of great interest, but all could be accommodated within the standard theory. But he adds that the discovery of just once instance of any of the unknown transfers "would shake the whole intellectual basis of molecular biology, and it is for this reason that the central dogma is as important today as when it was first proposed" (Crick 1970: 563).

Mayr (1982) notes that the term *inheritance of acquired characters* does not fully capture the ideas around the putative effects of environment and use and disuse upon heritable traits. This is because "the belief usually also included the postulate of a modifiability of the genetic material by general climatic and other environmental conditions... or by nutrition directly, without peripheral (phenotypic) characters acting as intermediaries" (Mayr 1982: 687). Thus, the genetic material was *soft* or pliable under this view. The opposite view is that the genetic material is unchangeable (by these means) and *hard*.

It is important to be clear that while the central dogma finds no role for induced phenotypic variation in recoding DNA sequence, and thus commits to *hard* genetic inheritance, this is not a denial that further variation can be induced in the phenotype.

Genetic variation most certainly introduces phenotypic variation, but genetically constructed phenotypes can be subject to other processes that change them (Vogt et al. 2008). This point is in danger of being missed because of the genotype–phenotype distinction originating with Johannsen (Johannsen 1911). Initially, under Mendel, external traits were understood merely as markers of internal units of heredity, but later the distinction became one between the potential for a trait and the actual trait (Falk 1986). The genotype was the potential, and it expressed the phenotype. According to some scholars (Falk 2010; Sultan 2019) this led to biologists regarding the phenotype as directly genetically determined. Although this is theoretically conceivable for some monogenic traits, we now understand most traits to be polygenic, and we also understand the role of gene expression to be hugely mediated by many factors external to core DNA sequences (Keller 2000). Regulation has been taken up by the evo-devo community, demonstrating how selection enables novel regulatory genes to direct conserved and compartmentalized developmental processes toward novel phenotypes (Gerhart and Kirschner 2007; Kirschner 2013; Kirschner and Gerhart 2010).

The new science emerging within evo-devo is not a challenge to the central dogma because it is not suggesting that any of the unknown transfers are occurring. Nor is it a challenge to the Modern Synthesis because it is simply filling in the details of development that were bracketed during its formation. More precisely, as discussed in Chap. 6, the theoretical constraints imposed upon what development should look like by the Modern Synthesis appear to have been upheld. However, since the 1990s, the molecular details of epigenetic effects have become better understood, and those advocating for an extension of the Modern Synthesis have claimed these discoveries as both vindication of certain kinds of Lamarckian or soft inheritance views and a refutation of the hardened synthesis (Chap. 3).

7.3 Epigenetics

The term *epigenetic* literally means *above the gene*. Waddington first used the term to refer to the causal mechanisms that allow the genotype to give rise to the phenotype, and he felt that this was a missing element in neo-Darwinian evolutionary theory.

> Because Waddington claimed to provide a richer paradigm for studying the interaction between organism and environment than the impoverished view of genetics, an epigenetic approach has appealed to critics of evolutionary "orthodoxy," whether these be biologists who feel that there is something lacking in the neo-Darwinian synthesis . . . or philosophers who favor a less gene-centric, more holistic, view of biology. (Haig 2004: 67)

This quotation focuses upon key elements of criticism of Modern Synthetic views, here aligned with neo-Darwinism in part due to the contributions of Weismann, I presume (Chap. 3). These criticisms are still very much at the heart of the contemporary criticisms, as we have already seen (Chap. 6). Haig goes on to provide a brief history of the somewhat varied use of epigenetics as a term, arriving at Nanney's substantial contribution.

> On the one hand, the maintenance of a "library of specificities," both expressed and unexpressed, is accomplished by a template replicating mechanism. On the other hand, auxiliary mechanisms with different principles of operation are involved in determining which specificities are to be expressed in any particular cell. Even without specifying precisely how these other mechanisms operate, the distinction between mechanisms involving template replication and "other mechanisms" is reasonably clear, even though both are involved in determining cellular characteristics. Difficulties arise, however, when one attempts to determine whether observed differences in cellular properties are due to differences in the "primary genetic material" or to differences in other cellular constituents. Some of these difficulties can be made apparent by setting forth certain general propositions related to the supplementary regulatory systems for which evidence is now available. To simplify the discussion of these two types of systems, they will be referred to as "genetic systems" and "epigenetic systems." The term "epigenetic" is chosen to emphasize the reliance of these systems on the genetic systems and to underscore their significance in developmental processes. (Nanney 1958: 712)

The library of specificities is Nanney's rather poetic description of the genotype. What Nanney is referencing is that each cell in a body contains the same DNA (except sex cells), but each cell performs a different function, and this is made possible through changes in gene expression. As an embryonic cell is pluripotent, to become specialized involves the silencing of genes for other functions, and this is achieved through various epigenetic processes. Thus the epigenetic control system determines what is expressed in each cell (Haig 2012). Nanney's view is quite different from that of Waddington, where the latter is firmly developmental and implies other options for the introduction of phenotypic variation beyond the switching between different functional portions of DNA. Both views developed separately for some time, but in the 1990s, DNA methylation was discovered and provided a plausible switching mechanism for gene expression. Moreover, the methylation state appeared to be heritable. This use of methylation was in keeping with Nanney's views on epigenetic control. By 2000 histone modification was also added to this view of epigenetics (see Sect. 7.3.1). But it was perhaps the heritable effects from the 1990s that were most influential.

> Holliday proposed that epigenetic defects in germ line cells could be inherited by offspring and proposed that 'heritable changes based on DNA modification should be designated epimutations to distinguish them from classical mutations.'

> For a brief period, epigenetics and DNA methylation became almost synonymous, at least in the Nanneyan tradition, with heritability recognized as a *condicio sine qua non* of epigenetics. Influential definitions from this period are 'Nuclear inheritance which is not based on differences in DNA sequence' and 'the study of mitotically and/or meiotically heritable changes in gene function that cannot be explained by changes in DNA sequence.' (Haig 2012: 15)

Haig (2012) notes the growing list of mechanisms added to the twenty-first century concepts of epigenetics, claiming that this has introduced some confusion and led to consternation by some critics, and this included suggestions that the term *epigenetic* should be abandoned (see (Jablonka and Lamb 2002) for an earlier discussion). Haig believes the term has persisted, in spite of significant changes in use, not least between Waddingtonian and Nanneyan traditions, because of its relation to the verb *epigenesis* and its classical meaning.

Pythagoras is associated with *preformation* while Aristotle is aligned with its opposite, *epigenesis* (Coall et al. 2014). Many preformation theories were homuncular, claiming that a small, but fully formed adult was harbored within the paternal or maternal substrate. *Preformationism*, then, is the nativist idea that all traits have a complete form at conception when male and female contributions meet. It is likely that those critics of the Modern Synthesis who claim that it had no role for development and saw form as genetically determined, also see what they call gene centrism as preformationist, or at least overly nativist (Dickins and Dickins 2018; Oyama 2000). Aristotle promoted epigenesis in which *material* causes were contributed by both parents; thereafter *efficient* causes led to the emergence of the *formal* cause, which has *final* purpose.

This view of epigenesis, something Waddington was happy to allude to (Haig 2004), combined with the etymology indicating it is above genetics has, according to Haig (2012), allowed disparate researchers to cohere under what he clearly considers an illusory banner. But Jablonka and Lamb, at the start of the twenty-first century, saw a core role the new epigenetics, that had emerged since 1990, within evolutionary theory:

> We can say that epigenetics requires a broadening of the concept of heredity and the recognition that natural selection acts on several different types of heritable variation. Although the current gene-centered version of Darwinism—neo-Darwinism—is incompatible with Lamarckism, Darwinism is not. In the past, Lamarckism and Darwinism were not always seen as alternatives: they were recognized as being perfectly compatible and complementary. In the light of epigenetics, they still are. Recognizing the role of epigenetic systems in evolution will allow a more comprehensive and powerful Darwinian theory to be constructed, one that integrates development and evolution more closely. (Jablonka and Lamb 2002: 95)

There are, then, at least two possible routes to take when adopting epigenetic work. First, one might take the Waddington tradition and argue for developmental processes between the genotype and the phenotype. This fits well the evolutionary developmental biology approach and is not necessarily in opposition to the Modern Synthesis, as it could be seen as providing the missing account of development. The second approach is focused upon both the heritable aspects of changes to gene function due to epigenetic processes and what that means for phenotypic variation. It is this second view that Jablonka and Lamb are championing above, and they are hinting that epigenetics might soften views of inheritance.

7.3.1 A Little Detail

Gene expression relies on transcription. The sequence of base pairs from a single strand of DNA is copied to a single-stranded mRNA (messenger RNA) molecule. Once completed, this mRNA molecule leaves the cell nucleus to connect with ribosomes in the cytoplasm. Ribosomes *read* the mRNA, three bases at a time and tRNA (transfer RNA) molecules bring specific amino acids to the ribosomes. The amino acids form the polypeptide chain, which must fold itself up to form a protein. This process is highly regulated by *transcription factors*, which are themselves proteins that repress or activate gene transcription. These proteins bind with regulatory portions of the genes (see Pabo and Sauer 1992) and are influenced by environmental data. The epigenetics developed in the 1990s onward focuses upon this regulation.

Genes have promoter regions that contain transcription factors and therefore must be accessible. But access is restricted or limited by the tight coiling of DNA around *histone* proteins, rendering inaccessible genes inactive until acetylation, phosphorylation, or methylation chemically change the nature of their bond to the DNA. This process is reversible and is a key mechanism in gene regulation. Histone proteins themselves can be epigenetically modified, leading to different DNA binding around them, in turn altering DNA expression. This is regarded as a post-translational epigenetic effect (Richards 2006).

Another mechanism for regulating transcription is through the addition of a methyl group to cytosine bases on the DNA. It is this mechanism that silences genes. This methylation of cytosine occurs at sites where a cytosine nucleotide is followed by guanine (CpG or CG sites), and this can prevent the binding of transcription factors. Genomic imprinting, X-chromosome inactivation and gene expression all involve this kind of DNA methylation.

Additional epigenetic mechanisms include noncoding RNA molecules, such as microRNAs (miRNA), which impact upon post-translational gene expression, by inhibiting protein translation, targeted gene deletion and gene amplification (Jablonka and Lamb 2008).

7.3.2 The Epigenetic Challenge to the Modern Synthesis

Jablonka and Lamb have laid out the epigenetic challenge to the Modern Synthesis very clearly (Jablonka and Lamb 2008). They begin by listing some of the key assumptions, as they see them, of the Modern Synthesis. These are:

1. Heredity is delivered by the transmission of germ line genes.
2. Hereditary variation is understood as variation in DNA sequence.
3. Hereditary variation is caused by sexual recombination and mutations, and it is not affected by the development of the individual.
4. Selection for the most part targets individuals.

5. Heritable variation has generally small effects and evolution is, for the most part, gradual, and macromutation is mostly continuous with microevolution.
6. Evolution is vertical descent with modification and horizontal gene transfer does not influence the branching structure of evolution.

They then list the challenges:

1. Heredity involves more than DNA.
2. Soft inheritance exists.
3. Given symbiont and parasite presence in hosts it may be important to regard communities as the target of selection and not the individual.
4. Saltation is common causing evolution beyond the species level.
5. A webbed pattern is more appropriate than a tree for incorporating hybridization, symbiosis, parasitism, and horizontal gene transfer.

Jablonka and Lamb directly align epigenetic inheritance with soft inheritance and note that the term has been used in two overlapping ways.

> The first, which we refer to as epigenetic inheritance in the broad sense, is the inheritance of any developmental variations that do not stem from differences in DNA sequence or persistent inducing signals in the present environment. This includes cellular inheritance through the germline, and soma-to-soma information-transfer that bypasses the germline: for example, soma-to-soma transmission through developmental interactions between mother and embryo. The second is cellular epigenetic inheritance, which is the transmission from mother cell to daughter cell of variations that are not the result of DNA differences. It occurs during cell division in prokaryotes and mitotic cell division in the soma of eukaryotes. Sometimes it also occurs during the meiotic divisions in the germline that give rise to sperm or eggs, so offspring may inherit epigenetic variations through the germline. The mechanisms that lead to cellular epigenetic inheritance can also lead to cell memory - to the persistence of functional and structural cellular states in non-dividing cells, such as most neurons. For example, it seems that early maternal behavior in rats has long-term effects that are associated with chromatin marks . . . in a key gene in their offspring's brain cells; changes in marks are also known to be associated with fear conditioning in rats. (Jablonka and Lamb 2008: 390)

They discuss four kinds of epigenetic inheritance systems, a categorization developed in previous work (Jablonka and Lamb 2006, 2007). First are *self-sustaining feedback loops* where gene products act as direct or indirect regulators for transcription. These gene products can be inherited by daughter cells during cell division leading to the inheritance of the same regulatory state, and Jablonka and Lamb note the importance of this process in development. This internal transmission of state relevant data is not a typical interpretation of inheritance, perhaps due to the focus upon the individual level that Jablonka and Lamb have already flagged. But by noting its relevance to development, they are, in essence, committing to the same view of individuals as that used in the Modern Synthesis at this point.

Their second system is that of *structural inheritance*. This is a template inheritance idea, where structures act as templates for building similar structures in daughter cells. Jablonka and Lamb reference prions as an example. Prions have been cited as evidence of the violation of the central dogma of molecular biology (Koonin 2012). Prion proteins in fibril formation can interact with other prions in soluble formation, inducing the latter to take on fibril formation and Koonin refers to this as a form of analog heredity, which is the same idea as the template metaphor of Jablonka and Lamb. In yeast, prions facilitate epigenetic inheritance of phenotypes, for example, by changing stop codon dynamics and thereby introducing novel proteins that can produce some beneficial new phenotypes. In yeast, this has been linked to a form of bet hedging strategy to deal with adverse environmental conditions. Koonin notes that prion-mediated epigenetic effects such as this can be genetically assimilated through the meiotic assortment of extant genetic variation. Irrespective of the nature of the mechanisms that deliver the assimilation, Koonin declares that this violates the central dogma as data is flowing from proteins to the genome. Of course, this is not what has happened. Instead, epigenetic marks have dysregulated stop codon functioning, which has necessarily revealed a different codon sequence, with impacts upon protein synthesis. Put more directly, the novel proteins that are formed are formed by the usual transfer of data from DNA → RNA, as discussed by Crick. It is just that more of the available data has been revealed. Critically, no new data has been created at the protein level and transferred to the DNA sequence as the sequence remains as it always was. Any subsequent genomes with different DNA coding, as a consequence of assimilation, are a consequence of selection, which is entirely in keeping with the Modern Synthesis.

Chromatin marks are the third system. They are

> (P)roteins and small chemical groups that are attached to DNA and influence gene activity. Relicts of these marks segregate with the DNA strands after replication, and nucleate the reconstruction of similar marks in daughter cells. Chromatin marks include modifiable histone and non-histone proteins that are non-covalently bound to DNA, as well as methyl groups that are covalently bound directly to the DNA. (Jablonka and Lamb 2008: 391)

The fourth system is *RNA-mediated inheritance* (Sect. 7.3.1).

Jablonka and Lamb argue that all four epigenetic inheritance systems are evidence for soft inheritance and play a role in evolutionary dynamics. For example, they discuss plant hybridization, noting a role for DNA methylation in silencing some genes leading to new variation over which selection can operate. As noted above, this is not really soft inheritance because the DNA sequence has not been altered, but rather the regulation of its expression has. This is a point about developmental processes, and those processes can introduce new selection dynamics due to changes in phenotype if they are stable population-level plastic responses to the environment. Those phenotypes might persist across limited generations as a result of directly inherited marks and environmental stability, and they may enable genetic assimilation as a result. As noted, any subsequent population changes in genotype are due to selection.

Danchin and colleagues describe the epigenetic challenge as a part of a broader set of discoveries that move the scientific community to look beyond DNA (Danchin et al. 2011). They remind us that natural selection operates at the level of the phenotype and claim that the Modern Synthesis has led to a focus solely upon DNA and the assumption that inherited phenotypic variation is genetically caused. As we have discussed, Mayr and others were very clear that selection operated over phenotypes and that there were developmental processes interceding between genotype and phenotype (Chaps. 4 and 6). But Danchin et al. wish to draw our attention not only to developmental processes, in keeping with the Waddington tradition, but to other, non-genetic ways of transmitting variation across generations, which will account for measured differences between individuals and contribute to population effects.

> Non-genetic inherited information can arise through several interacting mechanisms, including epigenetics, parental effects and ecological and cultural inheritance. All forms of genetic and non-genetic inheritance contribute to phenotypic resemblance between individuals. Distinguishing among these various components is crucial because their distinct properties affect evolutionary dynamics in different ways. In particular... accounting for the distinct properties of non-genetic inheritance may resolve some major evolutionary enigmas. (Danchin et al. 2011: 475)

This project is not dissimilar to the overall project from Jablonka and Lamb, in which they too posit symbolic and cultural inheritance processes alongside epigenetics (Jablonka and Lamb 2006). As with Jablonka and Lamb, Danchin et al. firmly align epigenetic processes with the potential to modify gene expression and link it to "post-translational modifications of histone proteins and methylation of cytosines in DNA" (2011, 476) as well as other gene regulatory mechanisms.

As previously noted, this kind of regulation may be a core aspect of development because it enables certain kinds of plasticity (Champagne 2013; Dickins and Dickins 2018; Morgan et al. 2005). One way to think about this is to see epigenetic processes as producing changes in the DNA context such that DNA data becomes differentially available for subsequent interactions (Chap. 5, Boisot and Canals 2004). So, if we understand DNA code as data being fed into a developmental context, that package becomes informational due to state changes in the emerging system (organism), in keeping with our previous discussion of information theory where *information = data + context*. We should expect epigenetic processes themselves to be sensitive to certain conditions, which they will treat as data. This does not violate the central dogma of molecular biology because the data transfer is from DNA → phenotype via proteins.

This view of epigenetics and development is not far removed from that of Nanney and the idea of an epigenetic control system, where the control system can be regarded in adaptationist terms (Dickins and Rahman 2012; Futuyma 2017). For example,

> Though there are many ways in which the concept of homology may be applied within the study of epigenetics and developmental plasticity, here I propose that epigenetic mechanisms can be conceptualized as a homology of process—a series of molecular changes involving protein–protein, protein–DNA, and enzymatic reactions that are a highly conserved biological strategy for allowing variation at the level of gene expression and

> ultimately in the phenotype of the organism in response to environmental experiences. From this perspective, it is the epigenetic mechanisms themselves (and the responsiveness of these mechanisms to environmental modulation) that are homologous across species and taxa. (Champagne 2013: 34)

Champagne's paper gives rich detail of this epigenetically induced plasticity. Included within her examples are parental effects. Parental effects can be understood as the influence of the parental environment on offspring, or more precisely, "as sustained influences on any component of the phenotype of the offspring that derives from a parent, apart from nuclear genes; thus there is an influence of parental phenotype on that of the offspring" (Kappeler and Meaney 2010, p.818). Parents are key environments for developing offspring, and sensitivity to parentally produced cues should not be surprising to evolutionary biologists concerned with robustness and plasticity.

Rodent models have provided good examples of parental effects. For example, stable and non-genetically heritable individual differences in the amount of maternal grooming have been demonstrated (Champagne 2008). Champagne argues that hypothalamic oxytocin neurons mediate a mesolimbic response, which in turn leads to behavioral differences between mothers. Mothers that engage in high levels of grooming undergo a steady increase in dopamine in the nucleus accumbens before grooming, but this does not happen in low grooming mothers. This low grooming effect is linked to low levels of oxytocin receptor binding in the medial preoptic area of the hypothalamus. The pups of these low grooming mothers exhibit a relatively prolonged passive stress response after a stressor has been removed in experimental settings. This is associated with behavioral effects such as reduced exploratory behavior and behavioral depression. These pups also display low estrogen sensitivity such that low grooming mothers cause a reduction in the level of estrogen receptors in the medial preoptic area of the hypothalamus, an effect linked to methylation in the promoter region of the estrogen receptor alpha gene.

With Rahman, I have previously interpreted these findings as evidence of a facultatively responsive set of mechanisms that enable pups to develop appropriate phenotypes to deal with environmental stress and specifically related this to the concept of reaction norms because this response is not invariant over time, but changes in the presence or absence of stressors (Dickins and Rahman 2012).

Parental effects are packaged as an inheritance because of their effect across generations. It is this transgenerational property that also permits ecological and cultural inheritance. This scheme is not designed to exclude genetic inheritance but has been claimed as a pluralist approach to the inheritance that incorporates processes that run parallel to genetic inheritance (Bonduriansky 2012). This reference to parallel processes makes clear that no one is claiming that epigenetic (or other) factors can in fact alter DNA sequence, but rather that multiple processes contribute to phenotypic variation and its persistence across generations.

Ecological parameters remain constant for some time across generations providing a consistent context for development, cultural parameters are learned, and parental effects induce epigenetic processes in offspring. While all of these are transgenerational factors that likely play a key contextual role in various aspects of the production of phenotypic variation, its stability across time and possibly its changes, they differ from genetic inheritance in that they do not conform to Weismann's distinctions. However, epigenetic marks can also be directly transmitted from parent to offspring.

> (T)ransgenerational epigenetic inheritance leads to the inheritance of epigenetic marks across generations. Although a small proportion of epigenetic marks seems to be transmitted to offspring in multicellular organisms, the transmission of epigenetic alterations of gene expression across generations has been demonstrated in numerous eukaryotes. Only transgenerational epigenetic inheritance can affect inclusive heritability. (Danchin et al. 2011: 476)

Genomic imprinting, the role of transposable elements and prions are all included in this category by Danchin et al. When considering the epigenetic inheritance of this sort, rather than induction, and we are bound to focus on Weismann's distinction between somatic and germline modification, which was a key transition to neo-Darwinism and the beginning of a shift into the Modern Synthesis.

> Two kinds of epigenetic modifications of germ-line DNA should be considered. The first kind consists of recurrent, reversible changes of epigenetic state (intergenerational switches). These can be considered analogous to the differentiation of alternative cell types by epigenetic mechanisms during somatic development. An example of such a switch is genomic imprinting, in which a gene's epigenetic state is determined by its sex of origin. In this case, a past environment (whether the gene was present in a male or female body in the previous generation) determines whether or not a gene is expressed in the current generation. What is 'best' for a gene when maternally-derived is sometimes different from what is best for the gene when paternally-derived. Genomic imprinting allows genes to play conditional strategies that depend on whether they are in a maternal or paternal role. The switch to a maternal or paternal state is a directed non-random change that is acquired during an individual's life and inherited by its offspring. (Haig 2007: 420)

Haig notes that a standard Modern Synthesis[3] interpretation of this kind of switching would be that it is a property in part of the DNA sequence (the relevant genes are required and also the relevant site for modification). Furthermore, natural selection would eliminate deleterious switching, and switching does not allow change of the sort associated with cumulative evolution because the switching merely allows flip-flopping back and forth between states. Haig tells us that his preference is to regard such states as *transgenerational ontogeny*.

> The second kind of epigenetic modification of the germ-line consists of more-or-less permanent epimutations that, once having occurred, are transmitted faithfully to future generations until the occurrence of the next epimutation. Such epimutations are potentially a source of long-term evolutionary information and of phylogenetic transformation, but are conceptually little different from genetic mutations. The foundational models of

---

[3] Haig actually refers to neo-Darwinism instead of the Modern Synthesis, I suspect because of the underlying Weismannian connotations of this paper.

> neo-Darwinism are older than the discovery of the double helix. Therefore, it is hardly surprising that these models are essentially neutral as to whether a heritable change is brought about by a change in DNA sequence or a change in cytosine methylation. If heritable epimutations resulted in phenotypic diversification in a genetically homozygous population, then those individuals with epimutations that enhanced their adaptedness to the local environment would leave more descendants and the favored epimutations would necessarily increase in frequency. The challenge to neo-Darwinism comes, not from the existence of epigenetic inheritance, but from the possibility that epimutations are directed rather than random. (Haig 2007: 421)

In spite of this comment about the neutral stance from Darwinism on the precise nature of heritable material, Haig goes on to point out that epigenetic processes are unstable (high epimutation rates) relative to DNA and thus have low fidelity, making the process of evolution by natural selection less likely if reliant upon epigenetic inheritance. Added to this, Haig makes clear that DNA is data-dense and thus can play a much more detailed informational role in biological processes relevant to evolution and development.

In closing his paper, Haig deals with the challenge to Darwinism noted in the last quotation above. Epigenetic inheritance, and also parental effects produce entirely non-random phenotypic results through interaction with DNA code. Such a stable, population-level outcome can only be due to evolution through natural selection, and he pushes a strong adaptationist line. He makes the interesting comment that this view does not rule out some interpretations from Lamarckian thought. Thus, he can envisage selection, building a mechanism that detects increased musculature in male arms and sends a signal to the sex cells that in turn builds bigger arms in male offspring. This could be achieved theoretically via some kind of epigenetic mark and passed across generations. But the muscle detection system and the signal-response system will all have been selected for, and the DNA that the epigenetic marks operate over will have to enable the relevant modifications.

7.4 Inclusive Inheritance

In Sect. 7.3.2 I noted that Danchin and colleagues tied together parental effects and epigenetic, ecological and cultural inheritance under the banner of inclusive inheritance. This was deemed permissible due to the commonality of transgenerational effects across all categories. More recently, Danchin has made an argument based on information theory for a tighter conceptual linkage within inclusive inheritance (Danchin 2013). He lays out his claims as follows:

> I (i) unify all sources of phenotypic variations within the concept of information and define an information-driven approach to evolution; (ii) discuss a major overlooked methodological problem leading to confounding sources of variation; and (iii) illustrate how putting concepts of information at the heart of evolutionary approaches will affect the emerging inclusive evolutionary synthesis. (Danchin 2013: 352)

7.4.1 Avatars and Information

Danchin (2013) begins by relating reproduction to transgenerational information transfer. Biological information is then defined as any factor that affects the phenotype in a way that may or may not influence fitness (Wagner and Danchin 2010). Phenotypic variation captures informational variation between individuals that is caused by genetic and non-genetic contributions. Individuals do not have precisely the same version of information as one another. By implication, biological information is *for* the construction of phenotypes where *for* denotes some causal role in construction.

For Danchin the DNA code, in a particular gene, is an *avatar* of genetic information in the same way as a particular DVD is an avatar of a film, where the film is really the story, a functionally defined entity. We can say that information is encoded within DNA, and that is why DNA is a code or an avatar. From this Danchin argues that it is "the genetic information, not the DNA molecule *per se*, that is the target of natural selection" (2013, 352). This highlights the distinction between the mechanisms of DNA transmission or epigenetic transmission from the information they convey. It is the properties of the avatars that determine what Danchin terms the laws of inheritance. Here the biological information appears to be what is represented by the avatar, or rather the avatar is a means to represent information. The physical configurations of the avatar are in some capacity identical to the information, and the information has its causal effect via the avatar. But in principle, the same information could be conveyed through other physical means.

It is possible to separate the avatar from its information and make broad, lawlike statements about information transfer in the absence of knowledge about the mechanism of that transfer. Danchin points to the emergence of Mendelian genetics after 1900, but prior to the discovery of the structure of DNA in 1953 as a case in point. Once molecular genetics began then many more detailed discoveries about genetic inheritance were possible. Meanwhile, "in the absence of knowledge about the avatars of most nongenetic information, current nongenetic studies focus on informational dynamics to establish general laws about these systems to understand their impact on evolutionary dynamics" (Danchin 2013: 352).

7.4.2 Information as Out There or as Systemic Realization

Wagner and Danchin (2010) regard information as bound to detection. We could interpret this in the sense that facts about x only become information when detected. This interpretation is potentially compatible with the definition of information used in this book: *information* = *data* + *context*. Facts can be regarded as data or cues about some externality that can be inputted into a context and have an effect (Sect. 5.2, and below). Detection can encompass inputting into a context. But under this liberal interpretation, there remains a possible difference because Wagner and Danchin also see *facts as containing information to be extracted or detected*, rather than regarding information as a functional relationship between an input and a context. This allows them to invoke a distinction between undetected facts as potential information and detected facts as *realized* information.

Their use of *realized* could be interpreted in terms compatible with the view of information used in this book. But I am less certain that this can be sustained because Wagner and Danchin deliberately distance their definition from that of Shannon (Shannon 1948), claiming that what matters in evolution is the ability to accurately assess an environment and not the degree of uncertainty of the organism. "Detectable information is thus extracted from facts that may affect an organism's accuracy in appraising the environment" (Wagner and Danchin 2010: 207). It is not entirely clear whether this means that facts can be more or less accurately detected or more or less accurately deployed or both. But what this does seem to mean is that information is *out there* to be grasped, rather than information is *realized* by biological systems.

For Wagner and Danchin, then, realization simply means extracted and used. Under the view defended in this book, the term *realization* does different work. Critically, biological systems have design characteristics that enable particular inputs to cause a state change. It is the systematic nature of input-to-state-change relations that is informative, such that information is a functional relationship between input and system. Thus, it is the system that *realizes* information in the sense of *creates it through interaction* (cf. Oyama 2000). For example, the mechanisms of protein synthesis take inputs from DNA, mediated by RNA (and other molecules as we are discussing in this chapter), and that leads to the formation of polypeptide chains that fold. The regular state changes through this sequence of responses to inputs at each step are information, understood in terms of uncertainty. Polypeptide chains, as they form, can have different sequences of amino acids. There are 20 amino acids related to 64 codons (three nucleotide sequences) all of which add to the probabilistic limits of the next amino acid in a focal chain being, say, tyrosine. The arrival of tyrosine will have a surprisal value in the context of the possible amino acids that can form the chain (calculated logarithmically; see below). Here we have information as a functional relationship between an input and a system, between data and a context, and the operation of the system *realizing* that relationship.

Under the information as *out there* view, it is easy to consider inputs as containing all that is needed for function. Colloquially, information is often used to denote data, the input regarded without context, but when we talk about detecting information, or acquiring it, we inadvertently objectify it and change how it is deployed. Most importantly, for this discussion, information is seen representationally, and its representational properties are implicitly invoked in discussions of accuracy and likely role in evolution. The *systemic realization* view sees selection as operating over phenotypes and selecting those that are sensitive to data inputs that enable adapted fit. The data come to represent the external world by virtue of that selection, and only in the appropriate systemic context. The *systemic realization* view relies upon what Danchin sees as avatars and strongly claims that without physical mechanisms, there would be no information, for they make use of data distributions in the world.

Yockey has defended a role for Shannon-type information in biology and argued that the central dogma of molecular biology is both a biological and a mathematical fact (Yockey 2005). This is due to the property of surprisal. In Chap. 5 we discussed Cohen's interpretation of surprisal in terms of the *just-so-ness* of a message, standing out against the background of alternative configurations (Cohen 2000). The lower the probability of a particular input then the higher its surprisal value upon input to the system, and the greater its informational value. Shannon measured this in terms of bits using a logarithmic formula.

If we think of a fair coin being flipped, it will land on heads or tails in equal proportion. A coin is a binary device with two equiprobable symbols, and it therefore has $\log_2 2 = 1$ bit of information per symbol (Floridi 2010). This measure is perhaps more readily interpretable as the number of yes-no questions one would have to ask to ascertain the identity of the upside of a coin drawn at random from a sample of flipped coins (Gardner 2014). In this case, the question-and answer sequence, [<is it heads?>, <no>], provides complete understanding. Compare this with a die, with six equiprobable symbols, and each symbol has $\log_2 6 = 2.58$ bits of information. And finally, 64 equiprobable codons mean that each codon has $\log_2 64 = 6$ bits of information. .

Yockey used this metric to argue that information (data) cannot flow back and forth between systems unless they share the same Shannon surprisal values.[4] Thus, DNA → RNA and RNA → DNA are entirely possible as they both use a 4-letter code that delivers the same entropy values (via codons). But DNA → Protein can only go in one direction.

> Two alphabets are isomorphic, if and only if, they have the same Shannon entropy. The Shannon entropy of the DNA alphabet and the mRNA alphabet is $\log_2 64$. The Shannon entropy of the proteome alphabet is $\log_2 20$; thus, like all codes between sequences that are not isomorphic, the genetic code has a Central Dogma. No code exists that allows information to be transferred from protein sequences to mRNA. Therefore, it is impossible for the origin of life to be "proteins first". (Yockey 2005: 21)

Following Yockey, using the 20 amino acids gives $\log_2 20 = 4.32$ bits of information per amino acid, compared to the 6 bits of information per codon. There is a fundamental informational asymmetry.

7.4.3 The Role of Information in Inclusive Inheritance

Danchin argues that genetic information is potential information that is extracted and used when "confronted by environmental information" (2013: 352). Both forms of information capture variation and contribute to the total phenotypic variation seen in an organism. As environmental information is composed of epigenetic, parental, cultural, and ecological information transmitted across generations, then its interaction with genetic information condenses the history of recent and more distal

[4] Surprisal values are more commonly referred to as Shannon entropy, but the terms are interchangeable.

ancestors. Of course, for both genetic and environmental facts to be used in this way, some form of detection process or mechanism is required, following Wagner and Danchin (2010). The implication is that this is both an evolutionary and a developmental process, and Danchin sees this use of information as a way to reconcile both disciplines.

This is a reconciliation sought by focusing upon the emergence of form and variation in form, and by not imposing any kind of structure on informational sources, processing, or integration. The desire is to give genetic and environmental information causal parity in the emergence of form. In many ways, this is an informational method of describing *gene x environment* interactions and also a suggestion about how to think about the missing heritability commonly found in genome wide association studies (although much of that might be due to unknown genetics (Plomin and Simpson 2013)). It is also a caution against assuming all inheritance is genetic and all variation is genetic. This is a clear stance against what has been labelled gene centrism.

It is also a claim for a useful framework:

> (F)or information that varies too rapidly relative to generation time . . . there is no selection for any transmission to the next generation. This can explain most non-transmitted phenotypic variation. . . At the other extreme, some environmental characteristics are so stable . . . that selection favors their encoding in a similarly rigid and mostly irreversible system to that of genetics. For factors that vary with intermediate rates of predictability, selection probably does not favor genetic encoding, but rather their transmission through reversible systems as this would enable the tracking of environmental change, a possibility that genetic encoding would hardly allow. Cultural inheritance, for instance, may enable the inheritance of information that is stable over several generations, but in a way that is sufficiently plastic to allow culturally inherited information potentially to track environmental changes. There is thus no selection for encoding such information more rigidly. Finally, transgenerational epigenetic inheritance . . . may encode even more stable, but still not totally stable, environmental information in a way that may be more rigid than culture, but that is still reversible. (Danchin 2013: 354)

In this passage, Danchin again writes as if the information is something out there to be grasped by the organism. Information is seen as facts about something, and therefore has intrinsic semantic content (or meaning) that can be more or less accurately detected and deployed (as noted Danchin's accuracy point is not entirely clear). Danchin's use of potential and realized information forces the *information-out-there* interpretation and makes the organism's role one of agency in capturing information. This is in stark contrast to the *systemic realization* view of the information I have advocated. Nonetheless, Danchin's agency boils down to specific mechanisms that track his form of information in terms of the rate-of-change categories.

The rate-of-change categories and their associated mechanisms could just as easily be derived from other information models. These observations are not intrinsic to his particular commitments here. Where Danchin does make novel claims are the places he discusses the properties of avatars.

> The properties of avatars of biological information also affect information dynamics. On one side, genetic information is encoded digitally in the DNA sequence. The epigenetic code is less explicit, partly because it has several avatars. Even when considering only methylation patterns, it seems that it is not the actual position of every methyl radical on the DNA molecule but rather the rate of methylation of a given part of a gene that regulates its expression. (Danchin 2013: 355)

Because Danchin has eschewed Shannon's concept of uncertainty, he has also abandoned the notion of surprisal and a method of measuring information. His preferred route to measurement is accuracy, which he ties to fitness outcomes, but accuracy can be delivered by Shannon-type systems that grade certainty with accumulated inputs and construct probabilistic architectures that can be state dependent. In simple terms, the likelihood of a particular input, x, changing the state of the system from $S_1 \rightarrow S_2$, can vary because of other inputs or the overall state of the system, for example, the amount of available energy it has. Such likelihoods will be mediated by systems that have a functional interaction with data and are organized in some kind of interactive structure (Eronen 2015). This multiplicity of state dependent and probabilistic mechanisms has the potential for error and thus variance in accuracy. We should anticipate selection will act to reduce error rates.

Danchin gives no citation for his comment about methylation rates, but a model from Jeltsch and Jurkowska discusses rates of methylation as a function of local epigenetic and enzymatic effects at specific sites of methylation on DNA (Jeltsch and Jurkowska 2014). Methylation rates, so controlled, enable *de novo* methylation and provide a model for how methylation might be stabilized. They sum their position in the abstract as follows:

> DNA methylation can be described by a dynamic stochastic model, in which DNA methylation at each site is determined by the local activity of DNA methyltransferases…, DNA demethylases, and the DNA replication rate. Through the targeting and regulation of these enzymes, DNA methylation is controlled by the network of chromatin marks. (Jeltsch and Jurkowska 2014: 310)

This kind of *epi-epigenetic* control system is a natural step from Nanney's view of epigenetics and conforms to the Shannon-derived view of multiple inputs and interacting mechanisms. In truth, Jeltsch and Jurkowska describe a broad mechanism that does work across aspects of methylation. This mechanism can be more or less accurate in its delivery, but one would anticipate selection on this to reduce error. There is every reason to believe that this selection, more generally for epigenetic processes, treats epigenetics as a phenotype underpinned by a genotype, such that the epigenome delivers plasticity. This is brought home by studies of the heritability of the epigenome that demonstrate greater epigenetic similarity in monozygotic twins than in dizygotic and greater discordance between non-relatives (Ollikainen et al. 2010). Put simply, this implies that heritable DNA variants afford specific kinds of epigenetic marks, such that a portion of epigenetically induced phenotypic variation is due to that genetic variation.

> For the other nongenetic inheritance systems, at present there is no clear indication of the existence of a specific avatar. Cultural information is probably stored in the brain, ecological information resides in the properties of the environment itself, and some parental effects involve hormones or antibodies, but it is more difficult to consider these as real codes. We can only talk of a code when there is an avatar. It may well be that some non-genetically inherited information does not involve a real avatar. (Danchin 2013: 355)

In this final quote from Danchin about the role of avatars, we see his information-as-out-there definition putting information into individual brains for cultural detection, into the environment for organismic detection, and he is uncertain about the role of hormones and antibodies because he cannot see them as codes. He is re-emphasizing his view of avatars as those mechanisms that encode information, the mechanisms that deliver information. But he states that hormones and antibodies are not codes and therefore not avatars of information, and yet he claims information is passed. Previously he noted that information dynamics could be studied in the absence of knowledge about avatars but under the assumption that they will be discovered. Given this, it is hard to understand what he means to claim here, but the strong implication is that as information is out there, it can be grabbed in other ways. If he were to adopt the Shannon-type view of *information* = *data* + *context*, he could immediately clarify matters by simply discussing hormones and antibodies as data inputs into systems, as contexts, that respond to them (cf. Cohen 2000).

Danchin concludes:

> (N)ongenetic inheritance probably does not affect the overall structure of the evolutionary equation, but it certainly affects the structure and, thus, the value of each of its three main components; heredity, variation, and selection. First, it is now becoming accepted that heredity incorporates nongenetic information. Second, previous studies have shown that mutation frequency can be greatly increased in methylated genes, suggesting that nongenetic inheritance also strongly affects variation. (Danchin 2013: 357)

This is a claim to general Darwinism (Lewontin 1970; Webb 2011) that is neutral with regard to the material of inheritance (Haig 2007). But it is also a claim that extra-genetic processes can affect mutation rates and thus increase the amount of novel variation that may be available for selection.

Epigenetic variation can be introduced by the stochastic epigenetic marking of the DNA code. In bacterial populations, this leads to phase variation (flip-flopping between phenotypes) in order to deal with rapidly changeable environments and is effectively a form of epigenetically induced bet hedging to improve robustness often in the context of host immune response, therefore facilitating virulence (Van Der Woude and Bäumler 2004). It is most likely an adapted response that enables change where mutation rates cannot. Here the epigenome is providing parametrized variation by regulating gene expression, albeit stochastically.

The reference Danchin gives to support this second point about epigenetics inducing mutation is a review paper assaying the role of epigenetics in cancer that discusses two possible routes for DNA methylation to increase cancer by causing "aberrant cell cycle control" (Gonzalgo and Jones 1997: 107) More recent articles, covering the role of epigenetics in cancer, have also emphasized the role of hypermethylation of tumor suppressor genes leading to their silencing and the opportunity for mutations to then occur (Brower 2011; Jones and Baylin 2002). These are not opportunities for positive selection but rather evidence of dysregulation. Danchin's adoption of this example is best described as a hopeful monster argument in that it is perhaps a nod to the developmental macromutations proposed by Goldschmidt in the first half of the twentieth century (Dietrich 2003). The hope might be that dysregulation could lead to useful mutation in developmentally important genes and provide novel phenotypes.[5] Hopeful monster arguments have been discussed within evo-devo to some extent, but Danchin does not explicitly make this link (see (Theißen 2006).

7.5 Conclusion

The *systemic realization* position makes the strong claim that without physical mechanisms, there would be no information. As evolutionary theorists, we should expect biological systems to make use of inputs that enable them to fit into their environment. Organisms should be composed of numerous systems that take external inputs, that in turn enable adapted functions. To this end, biological life generates information, and evolutionary processes organize the systems that do this.

The dysregulation of tumor suppressor genes by hypermethylation is due to inputs that cause local functionality in terms of stimulating hypermethylation, but they introduce chaos into the overall system designed to control irregular and exuberant growth. This is not systematic state change, even though we might colloquially talk about the systemic spread of cancer in a body. Cancerous cells escape the normal regulatory controls and defy developmental patterning. This kind of dysregulation is evidence of the fundamental nature of *data + context* relations in biology.

All the mechanisms and processes that Danchin has referenced in his model of inclusive inheritance—genetic, epigenetic, parental, ecological, and cultural effects—can be reinterpreted in light of the systemic realization framework. To that end, readers may wonder why I am not happy to adopt an inclusive inheritance view, as an aspect of general Darwinian thinking where that means neutrality about the mechanisms of inheritance and variation. Surely Danchin and I can sort out our differences about information later?

I am content to adopt inclusive inheritance in its *weak sense*. That weak sense simply means acknowledging that epigenetic marks can be transferred across a limited number of generations, that parents can cause phenotypic effects in their offspring because they are an environment for their development, that ecologies remain constant in some dimensions across generations providing stability during

[5]This is not dissimilar to the error made by Koonin (Koonin 2012) when discussing prior disruption of stop codon functionality (Sect. 8.3.2).

development, and that some animals can socially learn and thus transmit data between generations. All of these phenomena conform to a definition of inheritance that relies only upon similarity across generations, although the kinds of entities that are similar differs between each mode of inheritance. Moreover, I am happy to acknowledge an ontogenetic role for these processes in phenotypic variation.

The *strong version* of inclusive inheritance has it that all these systems have the capability to introduce phenotypic variation that can change evolutionary dynamics in a similar fashion and that all modes of inheritance are equal with reference to the transmission dynamics of evolution. This I do not accept, for the straightforward reason that it fails to take account of the dependency relationships between each mode of inheritance and their systems. This failure is due to seeing information as out there, as containing all that is required to produce form, and as extractable and usable by different systems.

As an example, epigenetic processes are the first and foremost processes of gene regulation. They are essential to cell specialization but also deliver different phenotypic outcomes by switching on or off portions of DNA. Having different phenotypes associated with DNA has been made sense of in the reaction norm literature but also in the bet hedging literature. To achieve these effects, the epigenome relies upon DNA. As a regulator, the epigenome enables DNA code to either be entered or not into subsequent processes, subsequent systems where they cause a state change. The complexity of epigenetic action is vast, but this idealization usefully captures what is being done and helps to organize our understanding. That some epigenetic marks, once activated, can be passed across a limited number of generations is important and enables prepared offspring benefits that will undoubtedly impact inclusive fitness.

Cultural inheritance relies upon social learning processes that are in turn the product of neurological processes. Demonstrating a skill or telling somebody about something are all forms of communication which have as components cues, signs, signals, or symbols which correlate with relevant referents and act as data for the recipient of any communication (Dickins and Dickins 2001). Communication can be achieved between two individuals or within groups, and novel ideas and practices can be more rapidly spread within populations as a result. This communicative ability is an aspect of the phenotype and is a clear candidate adaptation. This argument stands for non-human species also. What is interesting about social learning is that an organism that has made a discovery or developed a new practice finds a way of denoting it, so it finds a way of creating a stimulus that will help others to coordinate their behavior toward the novel situation. Communication is a socially instantiated system that takes specific inputs leading to state change at the individual level that can be multiplied up to the population level. But social instantiation quite literally means some brains are acting as emitters of data and others as receivers. In this way, information is created or systemically realized but against the background of individual learning histories, introducing much noise.

Social learning, and thus cultural inheritance, is a social coordination solution. It cannot exist without individuals who can learn and communicate. Learning is an aspect of physiological plasticity, which is a robustness solution, and it is likely that the properties of neurons and brain organization that enable learning are products of genes, some of which will be under epigenetic regulation (Dickins 2021; Dickins and Rahman 2012; Ginsburg and Jablonka 2019). All of these systems are related in an approximate hierarchy in terms of the outcome, understood as a fully functioning individual organism. But the detail of the organization in reality will be more heterarchical, due to feedback, for example, with learning exposures strengthening synaptic weights and introducing biases into future learning (Eronen 2015). But while feedback is a key aspect of control and regulation, ultimately, there is a starting point to be found.

The Modern Synthetic view of genes provides this starting point, placing genes as a necessary causal condition for development but not a sufficient one. They must be regulated, and they must interact, they must respond to context, etc. But this starting point also enables high-fidelity transmission across generations and is thus necessary for evolution also. Without this high-fidelity transmission then, for example, epigenetic processes would be chaotic, and the patterned phase variation we see in bacteria would not be possible.[6] Moreover, genes enable epigenesis in its old sense. They are the initiating condition for life within the minimal context and enable the unfolding of subsequent context. In this sense, genes are procedurally early and sufficient developmental causes.

One could take away cultural inheritance, parental effects, and epigenetic inheritance and still have minimal life with just genetic inheritance. This could even occur in unstable environments with suitable mutation rates. This counterfactual thought makes apparent that all else in the inclusive package are solutions for robustness, are dependencies of a genetic precursor. It should also make clear that none of the systemically realized information at higher levels of the organization, such as in brains, could happen without those systems that are beneath. This is true all the way down.

References

AlQuraishi M (2019) End-to-end differentiable learning of protein structure. Cell Syst 8 (4).292–301.e3. https://doi.org/10.1016/j.cels.2019.03.006

Boisot M, Canals A (2004) Data, information and knowledge: Have we got it right? J Evol Econ 14 (1):43–67. https://doi.org/10.1007/s00191-003-0181-9

Bonduriansky R (2012) Rethinking heredity, again. Trends Ecol Evol 27(6):330–336. https://doi.org/10.1016/j.tree.2012.02.003

Brower V (2011) Unravelling the cancer code. Nature 471:s12–s13

[6] Where we might say epigenetic inheritance has high fidelity over limited generations, we should probably note that this is a property of the DNA that is being marked. Contrast this with the higher level processes in the inclusive inheritance package, such as social learning, where we should expect and indeed see low fidelity errorful transmission. Social learning is only ever an approximation that biases later individual learning toward a target. That individual learning is idiosyncratic. Loss of fidelity is a property of distance from core, high-fidelity genetic data.

Champagne FA (2008) Epigenetic mechanisms and the transgenerational effects of maternal care. Front Neuroendocrinol 29(3):386–397. https://doi.org/10.1016/j.yfrne.2008.03.003
Champagne FA (2013) Epigenetics and developmental plasticity across species. Dev Psychobiol 55 (1):33–41. https://doi.org/10.1002/dev.21036
Coall DA, Callan AC, Dickins TE, Chisholm JS (2014) Evolution and prenatal development: an evolutionary perspective. In: Lamb M (ed) Handbook of child psychology and developmental science: Vol. III. Cambridge University Press, Cambridge, pp 57–105
Cohen IR (2000) Tending Adam's garden: evolving the cognitive immune self. Academic Press, New York
Crick F (1958) On protein synthesis. Symp Soc Exp Biol 12:138–163
Crick F (1970) Central Dogma of molecular biology. Nature 227(5258):561–563. http://cs.brynmawr.edu/Courses/cs380/fall2012/CrickCentralDogma1970.pdf
Danchin É (2013) Avatars of information: towards an inclusive evolutionary synthesis. Trends Ecol Evol 28(6):351–358. https://doi.org/10.1016/j.tree.2013.02.010
Danchin É, Charmantier A, Champagne FA, Mesoudi A, Pujol B, Blanchet S (2011) Beyond DNA: integrating inclusive inheritance into an extended theory of evolution. Nature Reviews. Genetics 12(7):475–486. https://doi.org/10.1038/nrg3028
Dickins TE (2021) Ecological psychology, radical enactivism and behavior: an evolutionary perspective. Behav Philos 49:1–23
Dickins TE, Dickins DW (2001) Symbols, stimulus equivalence and the origins of language. Behav Philos 244:221–244. http://www.jstor.org/stable/10.2307/27759429
Dickins TE, Dickins BJA (2018) The extent of the modern synthesis: the foundational framework for evolutionary biology. In: Burggren W, Dubansky B (eds) Development and environment. Springer, New York, pp 155–176
Dickins TE, Rahman Q (2012) The extended evolutionary synthesis and the role of soft inheritance in evolution. Proc R Soc B Biol Sci 279(1740):2913–2921. https://doi.org/10.1098/rspb.2012.0273
Dietrich MR (2003) Richard Goldschmidt: Hopeful monsters and other "heresies". Nat Rev Genet 4 (1):68–74. https://doi.org/10.1038/nrg979
Eronen MI (2015) Levels of organization: a deflationary account. Biol Philos 30(1):39–58. https://doi.org/10.1007/s10539-014-9461-z
Falk R (1986) What is a gene? Stud Hist Philos Sci Part A 17(2):133–173
Falk R (2010) What is a gene?-Revisited. Stud Hist Philos Sci C Stud Hist Philos Biol Biomed Sci 41(4):396–406. https://doi.org/10.1016/j.shpsc.2010.10.014
Floridi L (2010) Information: a very short introduction. Oxford University Press
Futuyma DJ (2017) Evolutionary biology today and the call for an extended synthesis. Interface Focus 7(5):20160145. https://doi.org/10.1098/rsfs.2016.0145
Gardner M (2014) Community ecology: analytical methods using R and Excel. Pelagic Pub
Gerhart JC, Kirschner MW (2007) The theory of facilitated variation. Proc Natl Acad Sci USA. https://doi.org/10.1073/pnas.0701035104
Ginsburg S, Jablonka E (2019) The evolution of the sensitive soul: learning and the origins of consciousness. MIT Press, London
Gonzalgo ML, Jones PA (1997) Mutagenic and epigenetic effects of DNA methylation. Mutat Res 386(2):107–118. https://doi.org/10.1016/S1383-5742(96)00047-6
Gould SJ (2002) The structure of evolutionary theory. The Belknap Press of Harvard University Press
Haig D (2004) The (dual) origin of epigenetics. Cold Spring Harb Symp Quant Biol 69:67–70. https://doi.org/10.1101/sqb.2004.69.67
Haig D (2007) Weismann Rules! OK? Epigenetics and the Lamarckian temptation. Biol Philos 22 (3):415–428. https://doi.org/10.1007/s10539-006-9033-y
Haig D (2012) Commentary: the epidemiology of epigenetics. Int J Epidemiol 41(1):13–16. https://doi.org/10.1093/ije/dyr183

Jablonka E, Lamb MJ (2002) The changing concept of epigenetics. Ann N Y Acad Sci 981:82–96. https://doi.org/10.1111/j.1749-6632.2002.tb04913.x

Jablonka E, Lamb MJ (2006) Evolution in four dimensions: genetic, epigenetic, behavioral and symbolic variation in the history of life. MIT Press

Jablonka E, Lamb MJ (2007) Précis of evolution in four dimensions. Behav Brain Sci 30 (4):353–365; discusssion 365–89. https://doi.org/10.1017/S0140525X07002221

Jablonka E, Lamb MJ (2008) Soft inheritance: challenging the modern synthesis. Genet Mol Biol 31(2):389–395. https://doi.org/10.1590/S1415-47572008000300001

Jeltsch A, Jurkowska RZ (2014) New concepts in DNA methylation. Trends Biochem Sci 39 (7).310–318. https://doi.org/10.1016/j.tibs.2014.05.002

Johannsen W (1911) The genotype conception of heredity. Am Natural 531(XLV):129–159

Jones PA, Baylin SB (2002) The fundamental role of epigenetic events in cancer. Nat Rev Genet 3 (6):415–428. https://doi.org/10.1038/nrg816

Kappeler L, Meaney MJ (2010) Epigenetics and parental effects. BioEssays 32(9):818–827. https://doi.org/10.1002/bies.201000015

Keller EF (2000) The century of the gene. Harvard Univeristy Press

Kirschner MW (2013) Beyond Darwin: evolvability and the generation of novelty. BMC Biol 11 (1):110. https://doi.org/10.1186/1741-7007-11-110

Kirschner MW, Gerhart JC (2010) Facilitated variation. In: Pigliucci M, Müller GB (eds) Evolution: the extended synthesis. MIT Press, London, pp 253–280

Koonin EV (2012) Does the central dogma still stand? Biol Direct 7:1–7. https://doi.org/10.1186/1745-6150-7-27

Lewontin RC (1970) The units of selection. Annu Rev Ecol Syst 1(1970):1–18. http://www.jstor.org/stable/10.2307/2096764

Love AC (2017) Evo-devo and the structure(s) of evolutionary theory. In: Huneman P, Walsh DM (eds) Challenging the modern synthesis: adaptation, development, and inheritance. Oxford University Press, Oxford, pp 159–187

Mayr E (1982) The growth of biological thought: diversity, evolution and inheritance. Belknap Press

Morgan HD, Santos F, Green K, Dean W, Reik W (2005) Epigenetic reprogramming in mammals. Human Mol Genet, 14(SPEC. ISS. 1):47–58. https://doi.org/10.1093/hmg/ddi114

Nanney DL (1958) Epigenetic control systems. Zoology 44:712–717

Ollikainen M, Smith KR, Joo EJH, Ng HK, Andronikos R, Novakovic B, Aziz NKA, Carlin JB, Morley R, Saffery R, Craig JM (2010) DNA methylation analysis of multiple tissues from newborn twins reveals both genetic and intrauterine components to variation in the human neonatal epigenome. Hum Mol Genet 19(21):4176–4188. https://doi.org/10.1093/hmg/ddq336

Oyama S (2000) The ontogeny of information: developmental systems and evolution, 2nd edn. Duke University Press

Pabo CO, Sauer RT (1992) Transcription factors: structural families and principles of DNA recognition. Annu Rev Biochem 61:1053–1095. https://doi.org/10.1146/annurev.bi.61.070192.005201

Plomin R, Simpson MA (2013) The future of genomics for developmentalists. Dev Psychopathol 25(4 Pt 2):1263–1278. https://doi.org/10.1017/S0954579413000606

Richards E (2006) Inherited epigenetic variation: revisiting soft inheritance. Nat Rev Genet 7 (May):395–402. http://www.nature.com/nrg/journal/v7/n5/abs/nrg1834.html

Shannon CE (1948) A mathematical theory of communication. Bell Syst Tech J 27(3):379–423

Sultan SE (2019) Genotype-environment interaction and the unscripted reaction norm. In: Uller T, Laland KN (eds) Evolutionary causation: biological and philosophical reflections. MIT Press, London, pp 109–126

Theißen G (2006) The proper place of hopeful monsters in evolutionary biology. Theory Biosci 124 (3–4):349–369. https://doi.org/10.1016/j.thbio.2005.11.002

Van Der Woude M, Bäumler A (2004) Phase and antigenic variation in bacteria. Clin Microbiol Rev 17(3):581–611. https://doi.org/10.1128/CMR.17.3.581

Vogt G, Huber M, Thiemann M, van den Boogaart G, Schmitz OJ, Schubart CD (2008) Production of different phenotypes from the same genotype in the same environment by developmental variation. J Exp Biol 211(4):510–523. https://doi.org/10.1242/jeb.008755

Wagner RH, Danchin É (2010) A taxonomy of biological information. Oikos 119(2):203–209. https://doi.org/10.1111/j.1600-0706.2009.17315.x

Webb RH (2011) If evolution is the answer, what is the question? J Evol Psychol 9(2):91–107. https://doi.org/10.1556/JEP.9.2011.2.2

Yockey HP (2005) Information theory, evolution and the origin of life. Cambridge University Press, Cambridge https://doi.org/10.1016/S0020-0255(02)00173-1

Niche Construction Theory

8

Abstract

Niche construction theory is placed in its recent historical context, emerging from work challenging sociobiology in the early 1980s. While this initial work was broadly adaptationist, it rapidly took on the dialectical structures advocated by Lewontin and placed organisms as agents within an expanded view of selection. This argument led to a focus upon change in kinds of individual rather than within populations, but the introduction of developmental processes gave niche construction theory a blended variation and transformational perspective on evolution. It is not clear that niche construction theory can make any predictions or give any explanations that cannot be derived from standard evolutionary theory. In response to this niche construction theory has been declared a heuristic for deriving hypotheses differently. However, advocates of developmental niche construction theory have also claimed niche construction to be a distinct evolutionary process associated with a distinct and different view of evolution.

Keywords

Sociobiology · Gene centrism · Levels of evolution · Organism · Agent · Interaction · Externalism · Construction · Niche · Selection · Dialectical · Marxism · Development

8.1 Introduction

This chapter places niche construction theory in its historical context. It began with an attempt to counter the gene centrism of sociobiology but within an adaptationist framework, but quickly shifted to a more dialectical theoretical structure following critical commentary from Lewontin. In its initial formulation niche construction theory was focused upon the role of organisms as agents in changing selection pressures and the resultant adaptations from this interaction. But later niche

T. E. Dickins, *The Modern Synthesis*, Evolutionary Biology – New Perspectives on Its Development 4, https://doi.org/10.1007/978-3-030-86422-4_8

construction theory also imported arguments from the developmental challenge to the Modern Synthesis, leading to a blended variational and transformational theory of evolution.

In this chapter I argue niche construction theory fails to challenge the Modern Synthesis. There are two reasons for this. First, when niche construction theory relies upon selection it relies upon standard evolutionary theory. This means that niche construction theory amounts to no more than a redescription of evolution with a focus upon the detail of precisely the sources of selection pressure. Second, when niche construction theory incorporates development, it does this to challenge gene centrism and to demonstrate the effect of developmental causes on final form. However, this challenge conspicuously fails to deal with counterfactual arguments that would lead theorists to ask why developmental processes operate in the way they do. As I show, this is done to resolve an error in the selective niche construction story but in so doing it changes the definition of what evolution is.

8.1.1 The Pre-history of Niche Construction

One historical antecedent to the development of niche construction theory was the production of a multi-level model of evolution in response to the emergence of sociobiology (Plotkin and Odling-Smee 1981). Plotkin and Odling-Smee take aim at the gene centrism they saw at the heart of sociobiology. Their complaint is not that genes are selected during evolution but rather the assumption that genes have the goal of replication *sensu* Dawkins (Dawkins 1976). This they see as "reification, hypostatization, or simply genetic mysticism" (Plotkin and Odling-Smee 1981: 227). They claim that sociobiologists misapplied the adapted purpose of phenotypic traits to the level of the gene. While they understand that this application was a metaphor, they believe that it led to a distorted or erroneous biology in which genes are seen as the "motive force of evolution" (Plotkin and Odling-Smee 1981: 227).

Plotkin and Odling-Smee refer to sociobiology as the monolithic theory. For them it misses the huge problem of bridging the gap between genes and behavior. It is this bridging function that a multi-level theory is designed to serve. They begin to outline this theory by discussing adaptations, careful to distance themselves from pan-adaptationism in keeping with then current debates within the field (Gould and Lewontin 1979). Adaptations must fit the external world in some sense, and they characterize this in epistemic terms in which the formation of an adaptation is to some extent the encapsulation of information about the world.[1] Their question becomes one of whether there is one or there are multiple adaptation formation

[1]The concept of information used at this point is a representational one, in that it is *about* something. This kind of intentional-level view of representation is more common in psychology and philosophy of mind, following Brentano's influence on discussions of the mind (Huemer 2019). But it can be rendered in terms of either an *information-out-there* view, or a *systemic realization* view, as discussed in Chap. 7. Thus, information about the world is extracted and stored, or mechanisms are produced that process data from the external world in a systematic fashion. On

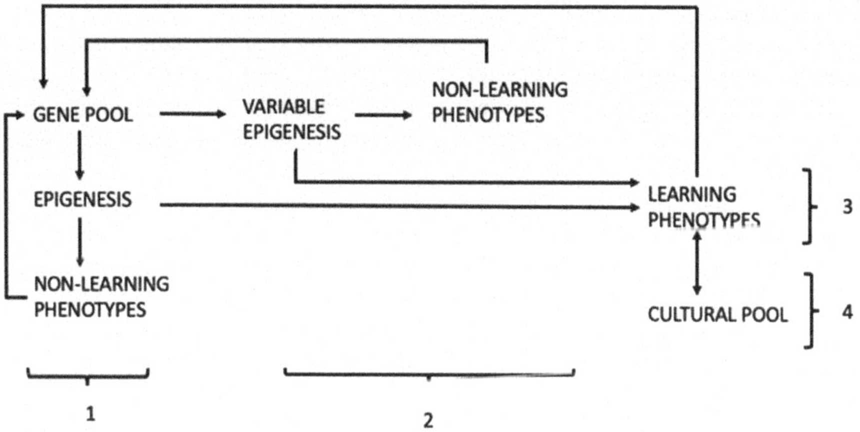

Fig. 8.1 Plotkin and Odling-Smee's four levels of evolution. The numbers refer to levels. 1 = Genetic, 2 = Epigenetic, 3 = Individual learning, 4 = Social learning. Arrows indicate putative informational integration between levels. Adapted from Plotkin and Odling-Smee (1981)

processes. They "believe that a complete evolutionary theory, that is, a theory capable of handling every known biological phenomenon, must be based on plural processes. Many adaptations can be formed by a single adaptation-supplying process, but there are some that require the operation of more than one such process" (Plotkin and Odling-Smee 1981: 228).

The multi-level theory is an adaptationist theory of levels of evolution. Each level is distinct from others and contains information, such that each level is termed a referent, and a referent is the beginning point of an adaptation. What this means is that information is moved out of storage and translated or transformed into an adapted phenotypic trait. A given adaptation may in fact rely upon information stored at multiple sites, and thus upon informational integration across referents. They give four levels of evolution, that are associated with different information (Fig. 8.1):

1. Genetic
2. Epigenetic
3. Individual learning
4. Social learning

These are not quite the terms that Plotkin and Odling-Smee use but they capture the distinctions. At level 1, the Genetic Level, the translational processes are in fact epigenetic, because of their role in gene regulation. Level 2, the Epigenetic Level, relies upon conditional epigenetic processes of the sort that allow phase variation. The authors explicitly reference Waddington's developmental view at this point, hence the use of *epigenetics*. Importantly, the authors make clear that the second

average, aboutness claims bias against a systemic realization view because it requires extra steps in reasoning.

level is nested under the first in a manner that imposes informational constraints from genetics upon epigenetics and that the epigenetic level is an adaptation that solves problems encountered at the first level.

Individual learning may or may not be nested under the second level, but it is always nested under the first. This imposes constraints on the third level also. Information gained at the epigenetic and individual learning levels is only retained by the individual which means that it is only the capacity to deliver outcomes, that yield success, that will be able to influence the gene pool. This is presumably because Plotkin and Odling-Smee are committing to the central dogma of molecular biology (Chap. 7). Furthermore, they possibly have in mind models of robustness in which parametrized developmental and behavioral plasticity are key categories of adaptation which they relate specifically to Mayr's discussion of open and closed programs (Chap. 5).

> The interrelations among these levels of evolution can be described by using a loose analogy with a computer algorithm. It is as if the main programme of evolution (what we call the primary level) has the potential to write supporting subroutines for itself if and when it starts to be overloaded by environmental change. These subroutines are the subordinate levels of evolution. The important point to note is that the evolution of the subsidiary levels is a response by a primary level that is in trouble, because its individual members are not coping with rapid rates of change in the world. This provides a clue for answering the question as to how, if a primary level has already evolved subsidiary referents, the demand for adaptation is shared across the multiple levels. This is in part an empirical matter, but we believe that we can supply a substantial guideline for the answer. Our basic hypothesis is that any environmental demand whose rates of change are sufficiently slow that it can be handled by the primary level, will be so handled. The subordinate evolutionary levels will come into operation only when the rates of environmental change lie outside the temporal operating range of the first level (Plotkin and Odling-Smee 1981: 230).

While the fourth level, or referent, of social learning clearly depends upon the ability to individually learn, and is thus nested under level one, Plotkin and Odling-Smee claim that social learning builds culture, which means that the information in that level is both stored within the nervous systems of all individuals within a culture and passed between them non-genetically. The authors see culture as similarly organized to the first referent but with its own potential evolutionary dynamic that is separable from the first level in a way in which levels two and three are not. Thus, there can be culturally created adaptations. Specifically, they define cultural evolution as Lamarckian.

> Cultural adaptations, like all other adaptations, are expressed by individual animals. However, as long as those animals continue to depend upon genes for their reproduction, they, along with all their adaptations, will continue to be subject to natural selection. Yet natural selection is indifferent to the source of any adaptation carried by an animal since it cannot select within animals but only between them. It is only sensitive to whether an animal as a whole is fit. Natural selection cannot, therefore, distinguish between a cultural and a noncultural adaptation within the same animal just because the cultural adaptation happens to have been transmitted nongenetically (Plotkin and Odling-Smee 1981: 232).

Plotkin and Odling-Smee are arguing that because natural selection operates over phenotypes natural selection cannot sift adaptations in terms of their underlying informational source. While all levels beneath the genetic have been claimed to be constrained by genetic information, at the fourth level the transmission of the adaptation can be direct and non-genetic, rather than indirect, via genetically endowed capacity. There is thus a degree of separation between culturally generated phenotypes and the underlying genotype. But because cultural learning could not happen without individual learning, we cannot decouple it from genetics. For Plotkin and Odling-Smee, cultural processes are just another example of robustness solutions, that enable rapid response to rapid change where mutation based evolution would not suffice. But note, they see the outcome of cultural learning as evolutionary. This is their innovation, to see shared knowledge as adaptations in their own right, rather than to see social learning only as the adaptation. In so doing they are committing to a form of general Darwinism, in that they see social transmission as inheritance and make an implicit claim to neutrality on the substrate of inheritance (Haig 2007; Webb 2011).

After making their arguments about culture, cultural adaptation, and the coupling of genotypes to cultural learning capacities, Plotkin and Odling-Smee make the case for an exception. They suggest that many species may be able, through their own actions, to modify the selection pressures they are under, and they think that socio-cultural processes may have a role to play in this. It is this idea that is at the heart of niche construction theory and within the decade the theory was being developed and it was taking direct aim at the Modern Synthesis for its reification of genes and its neglect of the impact of organismic action upon selection pressures (Odling-Smee 1988).

8.1.2 The Main Claims from Niche Construction Theory

In their book, *Niche Construction: The Neglected Process in Evolution*, Odling-Smee, Laland, and Feldman define niche construction as interactions by organisms with the environment that lead to the alteration of natural selection pressures for them and for others (Odling-Smee et al. 2003: 1). Interactions include resource extraction, habitat choice, the construction of artifacts, the emission of detritus, and death. They lay out four consequences of niche construction:

1. Ecosystem engineering
2. Modification of selection pressures
3. Ecological inheritance
4. Adaptation

Leaf cutter ants are used as examples of ecosystem engineers. Not only can they destroy plantations of particular species, but they can also aerate huge amounts of soil and circulate essential nutrients for other species. Where environmental modification to selection pressures occurs, that modification must be persistent for evolution to take hold. Odling-Smee et al. note the persistence can be delivered by niche-constructing genes that facilitate the constant repetition of specific behaviors or by the sheer magnitude of effort from ancestral generations that leads to change that lasts across generations. This is what the authors refer to as ecological inheritance and it "comprises whatever legacies of modified natural selection pressures are bequeathed by niche-constructing ancestral organisms to their descendants" (Odling-Smee et al. 2003: 13). Earthworms are also used as an example:

> Through their burrowing activities, their dragging organic material into the soil, their mixing it up with inorganic material, and their casting, which serves as the basis for microbial activity, earthworms dramatically change the structure and chemistry of the soils in which they live, often on a scale that exceeds even the soil perturbing activities of leaf-cutter ants. For instance, in temperate grasslands earthworms can consume up to 90 tons of soil per hectare per year. Similarly, as a result of their industry, earthworms affect ecosystems by contributing to soil genesis, to the stability of soil aggregates, to soil porosity, to soil aeration, and to soil drainage. Also, because their casts contain more organic carbon, nitrogen, and polysaccharides than the parent soil, earthworms can also affect plant growth by ensuring the rapid recycling of many plant nutrients. In return, earthworms probably benefit from the extra plant growth they induce by gaining an enhanced supply of plant litter. All of these effects typically depend on multiple generations of earthworm niche construction, leading only gradually to cumulative improvements in the soil... It is likely that some earthworm phenotypes, such as epidermis structure, or the amount of mucus secreted, coevolved with earthworm niche construction over many generations (Odling-Smee et al. 2003: 11).

Odling-Smee et al. argue that adaptation, like the phenotypic changes in earthworms, must be rethought, as a consequence of niche construction. They claim that standard theory overlooks the fact that "selective environments of organisms are themselves partly built by the niche-constructing activities of the organisms that they are selecting for" (Odling-Smee et al. 2003: 17). They tell us that this means the selection pressures cannot always be decoupled from the adaptation, and instead a system of feedbacks must be considered.

The decoupling argument follows work from Lewontin. Previously Lewontin had pulled Plotkin and Odling-Smee (1981) up on their lack of emphasis upon the role of agents' actions, implying that this should be foregrounded more in their work (Lewontin 1981). A couple of years later Lewontin discussed the relationship between organisms and the environment within evolutionary theory (Levins and Lewontin 1985).[2] He was particularly interested in accounts of the differential success of variants, and he noted that under one interpretation variants have differential reproductive success which will lead to one type becoming dominant in a population. This kinetic view, as he termed it, would not however account for the fit between organism and environment. For that, a second view was often adopted that saw the environment as setting problems for organisms, and those with the best

[2]This chapter on organisms and evolution was originally published as a single author paper by Lewontin in 1983 (Lewontin 1983).

morphology, physiology, or behavior had the best solution to those problems and would thrive relative to others.

Lewontin credited the popularity of this second view to both its appeal to engineering principles, such that organisms appeared designed, and the fact that many adaptations appeared to be convergent homologues. While he offers no citations either adopting or in support of this position, Lewontin is taking aim at the design talk of theorists such as Dawkins, and others pursuing the hardened synthesis. This is a part of Lewontin's general dissatisfaction with adaptationism (Gould and Lewontin 1979) and he doubtless found Plotkin and Odling-Smee (1981) to be too adaptationist in their tone.

Lewontin finds both conceptual and factual problems with the idea that the environment sets problems. First, Lewontin argues that there is an infinity of possible relationships between physical facts in the world and biotic levels of organization. He notes, for example, that there is no organism that "crawls on its stomach, lays eggs and eats grass, although snakes live in the grass" (Levins and Lewontin 1985: 98). He then asks if "evolution is now going on, as we assume it is, then what marks out the combinations of physical and biotic factors that makes the niches into which organisms are evolving?" (Levins and Lewontin 1985: 98). He further asks whether there is some form of natural kind, or a set of rules to be discovered that will answer this query. But he then points out that in the discipline of ecology a niche is defined by the activities, or what he calls practices, of the organism. He thus senses a contradiction between the view that the morphology, physiology, and behavior of an organism defines its niche and the idea that an organism evolves into a pre-existing niche. This is a contradiction that he believes is not resolved by classic Darwinian formulations of adaptation.

Lewontin's is a highly rhetorical argument in that he juxtaposes a before and after view of evolution in order to force the sense of a paradox, or contradiction where in fact he is simply referencing the creation of niches by evolutionary process. The rhetoric also relies upon the focus Lewontin draws to the individual, organism level while talking about evolutionary processes, which happen at the population level. It is true that we can look to an individual organism, as a member of a population, and discuss traits in terms of their adapted value to the organism but the individual organism did not evolve into a niche. Instead, evolution is the outcome of interactions between organisms and various externalities, which include both conspecifics, heterospecifics and physical environmental factors. Differential success is delivered between organisms by variation, and those variants that do better than others are replicated at a greater frequency across generations. The variation is produced randomly with respect to the outcome of selection and drift, but selection will pattern subsequent populations making it look *as if* individuals within it were deliberately designed to solve a contingent problem in the environment.[3]

[3]The reader will note that at this juncture I am being neutral with regard to the substrate of variation and inheritance, because this argument works at the general Darwinian level of explanation.

As Popper noted, the proper, nominalist use of definitions is as a short hand to capture a list of phenomena (Popper 1945). The term *evolution* should be applied in keeping with this stricture, such that *evolution* is not an essential kind in want of a definition, but rather the term we apply to the interactions and outcomes listed above. Given this it is important to only talk of populations evolving. Because the production and nature of variation is random with respect to evolutionary outcome there is no sense in which evolution occurs with reference to a specific aspect of the environment. But as Monod noted, when a variation is produced by chance, which enables relative success in an environmental context, selection will latch onto that (Monod 1971). This is easily interpreted *as if* there were some preordained necessity in the variation, when in fact it is a necessity of the statistical interaction that delivers evolution. Finally, describing a niche in terms of the traits of the organism that inhabits it is a legitimate way to define a specific niche, and is independent of the causes of this relationship. Indeed, a proper definition then demands an explanation, which is a secondary exercise. As my use of *as if* implies this secondary task can be greatly facilitated by hypothesizing that those traits might be adaptations (Dennett 1983, 1995).

Lewontin's factual concern is that of niche construction.

> The factual difficulty of formulating evolution as a process of adapting to preexistent problems is that the organism and the environment are not separately determined. The environment is not a structure imposed on living beings from the outside but is in fact a creation of those beings. The environment is not an autonomous process but a reflection of the biology of the species. Just as there is no organism without an environment, so there is no environment without an organism. The construction of environments by species has a number of well-known aspects that need to be incorporated into evolutionary theory (Levins and Lewontin 1985: 99).[4]

Lewontin then goes on to tell us that:

1. Organisms determine what is relevant.
2. Organisms alter the external world as they interact with it.
3. Organisms transduce the physical signals that reach them from the outside world.
4. Organisms transform the statistical pattern of environmental variation in the external world.
5. The organism–environment relation defines the traits selected.

[4]Lewontin's point can be made differently. The environment is that which is without the organism, whereas its ecology is that portion of the environment to which it is adapted. In systemic realization terms, this means that the ecology represents a set of data that can be inputted and treated informationally by the organism. I closed Chap. 7 noting that information is created by systems, and thus organic life creates information and evolutionary processes organize it. I think Lewontin's instinct is similar here, but he is wedded to a notion of agency and a colloquial view on information, no doubt.

What this all leads to is Lewontin's assertion that traditional evolutionary theory is one of organism as passive object, which should be replaced by a theory that incorporates organisms as constructors. He captures the distinction between passive and constructive theories in a set of equations that have been much replicated throughout the niche construction literature:

$$\frac{dO}{dt} = f(O, E) \tag{8.1}$$

$$\frac{dE}{dt} = g(E) \tag{8.2}$$

$$\frac{dO}{dt} = f(O, E) \text{ and}$$

$$\frac{dE}{dt} = g(O, E). \tag{8.3}$$

The first two equations are simultaneous differential equations that express the standard, passive organism view. Equation 8.1 captures the change *(d)* in an organism *(O)* over time *(t)* as a function *(f)* of the organism response to the environment *(E)*. Equation 8.2 is what Lewontin terms the law of autonomous change in the environment as a function *(g)* of some environmental variable *(E)*.

Equation 8.3 is a coupled differential equation showing coevolution of the organism–environment pair. Lewontin tells us that this coupled equation is hard to solve, but more importantly that it disrupts the neat cause–effect distinction of the traditional evolutionary view. Instead, there is a reciprocal process of selection for a trait that in turn affects the environment and then leads to selection of a modification and so on. It is for this reason also that advocates of niche construction theory, such as Laland, wish to move away from the proximate–ultimate distinction (Chap. 4, Laland et al. 2011). Lewontin offers this example:

> To understand the evolution of the sea lion from a primitive carnivore ancestor, we must suppose that at first the water was only a marginal habitat putting only marginal demands on the animal. A slight evolution of the animal to meet these demands made the aquatic environment a more significant part of the energetic expenditure of the proto-sea lion, so a shift in selective forces operated instantaneously on the shape of its limbs. Each change in the animal made the environment more aquatic, and each induced change in the environment led to further evolution of the animal (Levins and Lewontin 1985: 105).

Of course, all that this describes is gradual evolution over time, as the emerging sea lion species gains greater control over the aquatic environment. What Lewontin is rhetorically relying upon in this example is the idea that selection on the phenotype "made the environment more aquatic." It did not. It enabled the species to more fully utilize the aquatic environment. Sea lions are not making water.

The criticism of the sea lion example does not rule out the basic idea that organisms change selection pressures in their environment through modification.

The example of evolutionary transitions in earthworm secretions and epidermal structure in response to changes in their environment wrought by generations of ancestral earthworm burrowing activities is clearly important and of interest (Odling-Smee et al. 2003). But, the cause of environmental change, the origin of selection pressure, has no bearing upon the fundamental dynamics of evolutionary theory. For natural selection to occur one simply requires the basic ingredients of variation, differential success, and inheritance.

8.1.3 The Structure of Niche Construction Theory

In their 2003 book, Odling Smee, Laland, and Feldman end with a chapter advocating an extended evolutionary synthesis. Their claim is that niche construction is *not a product of evolution but an evolutionary process*, which is a co-contributor with, and by inference logically separable from natural selection. The co-contribution amounts to the introduction of a bias in natural selection as a result of perturbing natural selection pressures that might otherwise be in play (Heras-Escribano and de Jesus 2018). Thus, niche construction is not selection, but effectively generates specific selection histories, as we have seen above.

To make this case for an extended synthesis they borrow some explanatory distinctions drawn out by Godfrey-Smith.

> Externalist explanations explain internal properties of organic systems in terms of environmental properties. Internalist explanations are explanations of one set of internal properties in terms of another, and constructive explanations explain environmental properties in terms of organic properties... Adaptationism is generally externalist but adaptationists do not deny that internal factors such as genetic and developmental constraints exist and make some difference. It is a question of the relative weight associated with different explanatory factors (Godfrey-Smith 1996: 131–132).

Godfrey-Smith also defines asymmetric externalism that "(i) explains properties of an organic systems in terms of properties of the system's environment *and* (ii) explicitly or implicitly denies that these properties of the environment are to be explained in terms of other properties of the organic system" (1996: 132). Some commentators on niche construction theory have suggested that the Modern Synthesis, or rather standard views of evolution, are asymmetrically externalist in that they emphasize the causal role of external selection on genes in a way that privileges genes in explanations (see Aaby and Ramsey 2020). In effect, the argument is that the gene contains data (or more commonly information in most accounts) as a consequence of environmentally imposed selection and those data act as instructions for the phenotype. This then becomes the complaint against gene centrism and developmental plans, which as we have seen was an impetus for niche construction theory (Plotkin and Odling-Smee 1981).

Godfrey-Smith (1996) is careful not to accuse all standard evolutionary theory of asymmetric externalism when accounting for adaptations, as the quotation above makes clear. He points to cases, such as industrial melanism in moths, as good examples of the prioritizing of environmental influence on adaptation, and not without utility (Godfrey-Smith 1996: 136). But he is also clear that the introduction of game theoretical analyses, while externalist by his definitions, is not asymmetrically so. The impact of individual actions is central to those analyses (Maynard-Smith 1982).

Asymmetric externalism can fail to be true, under Godfrey-Smith's analysis, (i) when it is empirically the case that properties of an organic system are the consequence of internal dynamics and properties of the environment are a consequence of dynamics specific to the environment; (ii) when organic system properties are explained in terms of environmental properties but those environmental properties can be explained by properties of organic systems (an *interactionist* view); and, (iii) when internal dynamics of an organism contribute to properties of the environment (an *asymmetric constructivist* view). Niche construction theory aligns itself with interactionism (Odling-Smee et al. 2003, p. 373).

Externalities determine differential success in the Darwinian formulation. This fact leads to a very obvious question for niche construction theory—why are the actions and effects of organisms not externalities? Put another way—surely the effects wrought by organisms are a part of the environment? The dimension of time adds to this intuition (Uller and Helanterä 2019). Ancestral earthworms, burrowing and aerating soil, changed that environment in a way that provided novel selection pressure and change in subsequent generations. Selection had its effect in those subsequent generations because particular trait variants went to fixation at the expense of others. While this has an effect on the inclusive fitness of the ancestral burrowers, fitness gains, evolution, and the selection of adaptations require generational transitions. Of course, the descendants of those ancestors further changed the environment, leading to new selection pressures for their descendants, and so on—but this is clearly an organism-caused effect in the environment that leads to selection over subsequent generations. The only novelty here is the role of a minimal agent, in the sense of a behaving organism, and given Darwin's emphasis on the significant role of biota in generating selection pressure this is not really a novelty but rather a full rendition of his ideas. There is no teleology in this account: the ancestral earthworms were not designed to impose particular selection pressures upon their descendants. Those selection pressures that do emerge are accidental but consequential, in keeping with the idea of a bias. But that bias still relies upon the generation of random variation (but see below).

Lewontin's coupled equations (Eq. 8.3) have the potential to mislead, causing one to read the change in current organisms as determined by changes in the current environment that they have themselves wrought. But given that niche construction theory relies upon changes in gene frequencies, which in turn relies upon reproduction, a generational gap in this relationship is required. The O may change the current E, but the current E changes the future O, in Lewontin's notation.

Godfrey-Smith (1996) indirectly acknowledges some of the above issues:

> I have been directly contrasting internalist and externalist perspectives on organic systems, as if there is always a clear prior division between the system and its surrounds, between

> organism and environment. When working out whether an explanatory program is externalist it is best to allow the field itself to determine what counts as internal and what counts as external. Sometimes there is consensus about this, or an obvious boundary, but of course this is also an issue that is often contested. In fact a tactic use by some strongly internalist and externalist writers is to re-draw the boundary between system and environment in a way that makes the opposing view hard to sustain (Godfrey-Smith 1996: 48).

This confession leaves us in an odd position. What is the real debate here? Standard theory is happy to see actions such as burrowing and leaf-cutting, etc., as processes that have effects upon the world outside the organism, or agent delivering those behaviors. As discussed in Chap. 5, behavioral plasticity is a robustness solution under this perspective, and thus a solution to facilitate the maintenance of the genotype under change. But where those actions change aspects of the world, that are not aspects of the organism, standard theory has this as an environmental or external outcome. So, organisms behave in order to enhance their survival and reproduction in the course of a lifespan but there can be onward effects of those behaviors for others in the population that will, via natural selection, lead to population change. Those advocating for niche construction theory, do not focus on the individual in quite this way. They appear to see the individual as a *kind* of organism, with intrinsic and changing properties across time, and it is the actions of that kind and its slow *transformation* into a different kind that is captured by the *interactionist* approach they take from Lewontin.[5]

Odling Smee et al. (2003) reflect on the role of externalism in evolutionary theory, and quote Godfrey Smith again.

> For Lewontin, the imposition of an asymmetric externalist explanatory regime on biology was one of Darwin's central achievements. This enabled evolutionary theory to advance beyond (internalist) views in which changes undergone by species were understood as analogous to the unfolding of a developmental program. Darwin located the ordering mechanism of evolutionary change in the environment instead (Godfrey-Smith 1996: 141).

They use this quotation to emphasize the simplicity of standard theory and to ask what might be lost in a move to an extended interactionist theory. In highlighting this comment, they bring to the fore the tension between variational and transformational accounts of evolution. As suggested, the distinction between standard theory and niche construction theory may well be an attempt by the latter to retain transformational perspectives within the context of a variational theory. By shifting to discussion of *kinds* of individual, spread across generations, niche construction theorists can allow developmental processes back into evolution. If the focus remains upon natural selection organizing the fates of individuals as tokens, due to their relative

[5]Godfrey-Smith (1996) has Lewontin as a strong constructivist (pp.142–144) but his commitment to gene-level effects in population biology causes me to doubt this. Godfrey-Smith bases his claim upon the five points Lewontin lays out for incorporation into evolutionary theory (Sect. 8.1), but this is an argument against adaptationism, not against interaction between genes and environments via behavior. As such it flows from the general effort to avoid gene centrism.

success, leading to population level effects then development is excluded as an evolutionary process.

Uller and Helanterä hint at an interpretation of this sort and remind us that niche construction theory has its ancestry in arguments against gene centrism. They do this in order to draw a distinction within niche construction theory:

> When heredity is equated with transmission of genes, evolutionary change can be represented as transgenerational change in gene frequency. But if genes are not privileged as causes of heredity, phenotypic evolution may also occur through persistent changes to the developmental niche, that is, through epigenetic, social, and environmental mechanisms. This may raise concerns about the validity, or at least completeness, of evolutionary models whose currency is restricted to one out of possibly many sources of heredity, some of which may qualify formally as inheritance systems. Furthermore, a broader notion of inheritance appears to grant the possibility that development can direct evolutionary change through biased acquisition and transmission of features, which makes natural selection but one of several causes of adaptive change (Uller and Helanterä 2019: 358).

Here the authors invoke the notion of inclusive inheritance that Danchin has defended (Chap. 8, Danchin 2013; Danchin and Pocheville 2014) in order to contrast *developmental niche construction theory* (DNC) with *selective niche construction theory* (SNC) (Stotz 2017). The latter is the account of changes in selection pressure due to the actions of organisms that originated with Odling Smee and colleagues.

Developmental niche construction theory relies upon inclusive inheritance to account for the environmental fit of phenotypic traits. Different models of inheritance all require transmission, but the generational constraint is loosened under inclusive inheritance. For example, social transmission can occur within a generation allowing the spread of a learnt behavioral response to an environmental contingency. Equally, epigenetically mediated reaction norms can be stimulated by environmental exposures and could potentially have an effect upon the population presentation of a phenotypic trait. But, in keeping with some evo-devo arguments, the mechanisms of inclusive inheritance also enable the environment to produce novel variation, by creating a developmental niche (Stotz 2017).

> The relationship between SNC and DNC may best be compared with the relationship between the modern synthesis and evolutionary developmental biology. The latter provides the developmental mechanisms that connect the phenotype with the genotype, may account for the origin of variation, and highlight the effect of these variations on natural selection without directly affecting the construction of the selective niche. DNC spells out how developmental mechanisms, particularly the construction of a developmental niche, influence the origin of heritable variation and natural selection through the reproduction of a developmental system at the individual level. In other words, DNC is concerned with the origin of potentially adaptive, heritable, phenotypic variation. So while standard evolutionary theory assumes that all adaptations are the result of natural selection at the population level, this paper addresses the possibility that development can account for the creation of adaptations without invoking selection—a point that has been dubbed the arrival or origin of the fittest, rather than its survival (Stotz 2017: 2).

Uller and Helanterä are aware that all of the mechanisms and processes of inclusive inheritance can be interpreted in a way that makes them compatible with the Modern Synthesis, as I have attempted to show throughout this book. But Uller and Helanterä see these different interpretations as reliant upon different models of causation.

> A shared feature of the selective and developmental niche construction literature is that adaptive evolution is described as a reciprocally caused process. Natural selection and niche construction, the latter through its effects on variation, selection and inheritance, are concurrent processes that share responsibility for the complementary fit between organism and environment. In contrast, in the genetic theory of evolution natural selection alone is responsible for adaptation (Uller and Helanterä 2019: 362).

Reciprocal causation relies upon the reciprocal actions of proximate and ultimate causation, as outlined by Mayr (Dickins and Barton 2012; Mayr 1961), but in this quotation we see a claim for asymmetric externalism in the accounts of adaptation, where adaptation is the fit between organism and the environment. This is a profoundly Darwinian notion of adaptation, rather than a Modern Synthetic one as the latter would include discussion of gene frequency effects. Uller and Helanterä claim that Modern Synthesis chose to interpret the components of Darwinian evolution—variation, differental success, inheritance—as quasi-autonomous.

> To illustrate, imagine that three people are to paint a house. The first person goes to the store to buy the paint, the second mixes the paint, and the third does the painting. Further imagine that each person receives instructions separately and independently. Although the decisions that go into the buying and mixing of the paint affect the colour of the house, buying and mixing do not influence how the house is painted. The same applies to the three components of adaptive evolution in the Modern Synthesis. Variable rates of survival among individuals with different features determine what features will occur in the next generation. However, selection does not affect the process of inheritance; inheritance is merely the passing on of whatever genes were selected, typically following Mendelian rules. The variation that fuels evolution is similarly autonomous. Mutations occur randomly with respect to their consequences for development and fitness, and the acquisition of new variants does not change how variation is transmitted down generations. Each step determines (partly) the inputs for the next step, but not how those inputs will be processed. (Uller and Helanterä 2019: 363)

Accordingly the quasi-independence assumption makes only those processes that survive selection and inheritance count as evolutionary. Any trait that persists across generations will then become associated with genes, the Modern Synthetic candidate for stable inheritance and the developmental niche only becomes a stable context in which genes express themselves. Under this view, plasticity becomes incorporated as a property that is genetically conferred.

In order to counter this Modern Synthetic view Uller and Helanterä give an example of Beach Mice, in Florida. Their pale coat is due to a single nucleotide substitution that in ancestral environments may well have made them conspicuous, and accounts for the relative rarity of the type. But on sand dunes it carries an advantage and such mice have increased in frequency in these areas. This standard theoretical account relies on quasi-independence with random mutation and stable development being assumed in the background. Uller and Helanterä ask us to consider an alternative explanation in which "we may shift our evolutionary explanandum from the colour of the coat to the complementary fit between the mice's colour and their environment" (Uller and Helanterä 2019: 365). They claim that while the mice cannot change their color they can change how they behave by dispersing, or not, by becoming more risk averse or not. Each of these possible changes may well make the mutant coat color more efficacious in the dune context and will provide a developmental niche in which subsequent generations of pale mice adopt the interactive practices of their successful forebears. The "significance of this in terms of evolutionary causation is that the mechanisms by which individuals interact with their surroundings (that is, proximate causes), contribute to the directionality of evolution typically aggregated under the label natural selection" (Uller and Helanterä 2019: 365). Developmental causes become evolutionary in this scheme because they are recurrent, not because they are inherited in the sense typically associated with genes.

Uller and Helanterä end their paper with a discussion of the process of science and they align the emergence of niche construction theory with an attempt to broaden the phenomenal concerns of evolutionary theory in order to highlight different ways of describing the evolutionary process. This is evident in their conscious and explicit shift of explanatory target in the beach mice example. Nowhere do they deny that the pale morph persists due to selection on the single nucleotide substitution, and they clearly understand that it may do this due to a crypsis effect. But, what they want to add is that crypsis can be made more effective in fitness terms by behaviors such as dispersing to beach environments, and once in beach environments acting more boldly in order to gain resource relative to other non-crytpic competitors, etc. They are clear that these behaviors are unlikely to be linked to the coat color mutation. This enriches the story of the effectiveness of pale coat color on the dunes of Florida, but does it really change evolutionary accounts? A test of this, that Uller and Helanterä do not pursue, is the counterfactual. Would they expect pale mice to be selected in a beach environment, over darker morphs, even if the behavior of all mice were held constant. In other words, what if pale mice did not learn that they could be bolder and they did not learn that they should permanently disperse to beach environments, etc.? My assumption is that they would concede that basic predation dynamics would leave more pale mice and pale mice descendants. They would further concede that the addition of learning, etc., would change the rate of selection but not the qualitative outcome.

Under a standard theoretical account we might argue that a reduction in predation risk, due to crypsis, enables changes in fear response and an increase in boldness. We might also argue that mice track predation risk and move to avoid it and we can further argue that social learning around risk reduction makes good, adapted sense. In other words, we would ask why these processes are as they are and what are they for? All of the phenomena that Uller and Helanterä wish to consider are not excluded by the Modern Synthesis and standard robustness perspectives, but Uller and Helanterä do not ask these kinds of questions. Simply repackaging proximate causes, such as development, as evolutionary by arguing that they are recurrent is to change the definitional work done by the term *evolution*. Uller and Helanterä will object that this ignores their implied claim that niche construction theory sees all causes as interdependent upon one another, rather than quasi-autonomous. But, pale mice becoming emboldened under reduced predation risk in the context of beach environments is equally quasi-autonomous as the counterfactual demonstrates.

What developmental niche construction does achieve is the removal of the implicit metaphysical positioning around organisms as kinds stretched across generations that is a consequence of the original formulations of selective niche construction theory. Selective niche construction theory clearly conformed to the quasi-autonomy of standard theory and then made a muddle when trying to over-claim the role of the organism in its own evolution, a concept that is simply a non-sequitor under standard population level definition of evolution. Developmental niche construction theory avoids this by redefining inheritance, rather than evolution. The obvious problem with the definition based on developmental recurrence is that this requires a causal account that is truly separated from genetic inheritance. If there is any sense in which genes provide a necessary and prior condition for developmental processes to unfold, then recurrence is readily fitted back into standard theoretical interpretations.

8.1.4 The Constructions of Niche Construction Theory

The emphasis upon organisms as active agents in their own evolution is an outcome of Lewontin's commitments to dialectical modes of thought associated with philosophical Marxism and derived from Hegel (Levins and Lewontin 1985). As such this accounts for the strong historical character of niche construction theory, and in particular developmental niche construction theory and its emphasis upon a rich and detailed account of all aspects of the causation of an organism's fit to the environment. Marxist theory has a focus upon complex, historical transitions. Maynard Smith discusses Lewontin's commitment to dialectical materialism:

> The idea of a gene that influences development, but is itself unaltered, is undialectical—a view that I once held sufficiently seriously to spend some months on an (unsuccessful) experiment aimed at demonstrating the inheritance of an acquired character in Drosophila. This dialectical view has led many of those who have been influenced by Marxism to reject the idea that genes play a special role in evolution, and to be critical of the "gene-centered" approach pioneered by Hamilton, Williams, Dawkins, (and others). This reaches its strongest expression in the "developmental systems" approach, pioneered by Oyama, Gray, and Griffiths… The approach rejects the view that there is a fundamental difference between genes and other factors during development. After all, the cause of the difference in height between a typical and a dwarfed human could be that the latter carried the gene for achondroplasia, or had been seriously malnourished. Developmental system theorists propose a broader view of inheritance, which would include not only genes, but also such things as symbiotic microorganisms, social traditions, and aspects of the habitat (Maynard Smith 2001: 1496).

Maynard-Smith later notes that developmental systems theory "is related to Lewontin's idea that organisms should be seen as 'constructing' their environment" (2001: 1497), in that it is another move away from what Plotkin and Odling-Smee (1981) saw as the reification of the gene. Uller and Helanterä's emphasis upon quasi-autonomy, in the context of support for reciprocal causation, is also a plea for dialectical modes of thought and clearly closely aligned with developmental systems theory (Griffiths and Gray 2000, Chap. 9).

Lewontin was certainly a committed Marxist during the period in which he outlined his version of niche construction. In a 1982 paper, with colleagues, he described how biological determinism was used by the bourgeois classes to argue for intrinsic individual differences, and thus skills, that enabled each to find a natural level. Once the biases of the previous ruling class system were removed then any remaining inequality was naturally occurring, and the bourgeoisie were deserving of their elite status and disproportionate share of the means of production. It was a short step from this, the authors claimed, to making arguments from genetic determinism. But their central claim is that biology was constructed as deterministic to support the emerging bourgeois elite (Lewontin et al. 1982). For Lewontin the gene is not destiny, and an agent's own actions can determine its future.

These political views are not a necessary corollary of biological theories. It is perfectly possible to be a standard, Modern Synthetic evolutionary theorist and to seek equality in society. For example, by understanding variation within a population we have a greater chance of engineering our society in order to facilitate the potential of all. This is a view of our society as an environment that is not optimal for all, and a commitment to reconstructing that niche in order to approach optimality.[6]

It is also not the case that scientific views born of a political philosophy or commitment are necessarily wrong. As politics is about portions of the social ecology and the knowing uses of that ecology, it is perfectly possible that political descriptions will in turn capture something true about this type of biological functioning. In this context it is interesting to note Maynard Smith's nod to his own dialectical views, and how results led him to rethink those opinions. But Maynard Smith has also made the argument that Lewontin's Marxism has had little influence on his biology, other than his insistence upon the importance of context and interaction (Maynard Smith 1988). When discussing this, Maynard Smith makes comment on the role of mathematics in understanding complex systems and how this might replace the verbal arguments of dialectics and other Hegelian frameworks, which he takes as an attempt to understand complexity. He notes that a set of differential equations can be used for complex systems, and that their parameters can be gradually changed leading to gradual change in the system, but after a certain threshold the system will flip into a radically different state: he references a shift from gentle and gradually increasing oscillation to exponential growth. He regards this as the mathematical description of a quantity to quality transition, a

[6] I am indebted to Helena Cronin for pointing this out to me in discussion 15 years ago. She made the point much more elegantly than I, as she always does, and it stuck with me ever since.

non-dialectical transition also posited by Hegel although not widely taken up by all Marxists (Carneiro 2000).

It is not clear whether all or any of the various scholars committed to niche construction theory share the Marxist ideology of Lewontin. It is perfectly possible that they are adopting his position as one associated with downplaying genes and placing adaptation in a different context. As I have attempted to show in Sect. 8.1 the organism–environment coupling of niche construction theory is not a convincing counter to standard theory. Moreover, in some influential renditions the involvement of niche construction in challenging the Modern Synthesis is regarded more as the creation of a complementary approach which suggests that a form of pluralism is being advocated, rather than an analytic counter claim, even if the hope is that this will lead to new theory in the end (Laland et al. 2004; Odling-Smee et al. 2003; Uller and Helanterä 2019). In these cases it is often a mild claim that causal detail, for specific evolutionary accounts, are missed by the standard rendition, and this sounds like an issue of which idealization to favor in order to deal with causal complexity (Chap. 4, de Regt 2017; Potochnik 2020).

The systemic information theoretic perspective defended in this book can be regarded as an externalist theory, following Godfrey-Smith (1996). Data takes on an informational role within a context when it facilitates state change. I have been presenting biological systems as contexts that take data. Systems can change state in response to data in a way that enhances the fitness of the organisms that embody those systems. In this sense systems are internal to organisms, but responsive to externally derived data, where external is understood as relative to the system, not the organism. As those systems are a part of the phenotype, they are subject to natural selection and the account of the persistence of systems in a population (species) and changes in frequency of system types across time within a population are evolutionary accounts. This is a Darwinian perspective and provides a functional route to explaining the apparent purposive nature of biological life. The systems do not evolve, just as the organisms do not evolve, but variations in the systems may outcompete one another and lead to population dynamics and evolutionary change.

While systems may have been selected for, and thus explained in externalist terms, the use of the environment, external to the organism, is explained in terms of the internal, data sensitive properties of the system. Here use includes the categories of developmental and physiological plasticity, where the phenotype is regarded as adaptively responsive and much of that adaptive response is adapted. Those internal properties of organisms are explained with reference to genetics (an internal property) and the environment creating an *interactionist* explanatory framework. Given this, niche construction theory cannot lay claim to sole ownership of an interactionist perspective. But standard theory is not constructivist, in the sense that the focus of the Modern Synthesis is not upon the causes of change in the environment, but rather upon their effects. Where organisms interact with the environment it is in order to have some kind of fitness effect under the usual modelling assumptions.

8.1.5 Summary and Conclusion

Niche construction theory began life as a reflection upon different sources of adaptation by Plotkin and Odling-Smee (1981) as a challenge to gene centrism. In passing they noted the potential role of organisms, as agents, in the creation of novel selection pressures. This idea was compatible with Lewontin's philosophical Marxism and his emphasis upon dialectical modes of explanation, such that organism and environment should be seen as interacting partners in evolution. This idea has since formed the basis for niche construction theory as a theory of selection. Later niche construction theory embraced evolutionary developmental biology and concepts of inclusive inheritance to explain the apparent fit of form to the environment in developmental, transformational terms. By focusing upon changes in kinds of organism, rather than more formally on population level change, this enabled niche construction theorists to blend transformational and variational accounts in the hope of challenging the Modern Synthesis.

Some within niche construction theory have focused more upon the rich historical detail provided by theories conforming to dialectical structure. They have argued that this focus enables a fuller understanding of evolutionary dynamics by removing the quasi-autonomy criterion for components of evolution. But, as I hope to have shown, this strategy fails to deal with counterfactual possibilities and itself relies upon quasi-autonomy as all reciprocally causal arguments in fact do (Chap. 4): for there to be interaction there must be things that interact and are separably describable and causally separable. Moreover, the rich account view is compatible with game theoretic modelling developed during the final phase of the Modern Synthesis (Maynard Smith, 1982). What is interesting about this is Maynard Smith's reflections upon Lewontin's Marxism, himself a lapsed Marxist and pioneer of evolutionary game theory. This suggests that Maynard Smith found something explanatorily more satisfying in the Modern Synthesis than in his previous exposures to dialectical reasoning.

This last point raises an obvious question. *just what can niche construction theory do that is different from the Modern Synthesis?* With colleagues, I have asked this question (Scott-Phillips et al. 2014). This paper was unusual in that it contained an advocate of niche construction theory (Laland), and the other authors were skeptical of it. We asked whether new insights or predictions can be derived from niche construction theory (NCT) that cannot be derived from standard evolutionary theory, here referred to as neo-Darwinism and reached the following conclusions:

> We all agree that it is not logically necessary to use NCT to study and make scientific predictions about the natural world. NCT does not suggest: (i) that standard neo-Darwinism fails to recognize that organisms modify environmental states; (ii) that standard neo-Darwinism only considers selection pressures that emanate from the abiotic environment; or (iii) that it is not possible, even in principle, to provide explanatory accounts of biological phenomena with standard neo-Darwinism. In the light of this agreement, the skeptics see no reason to think that whatever predictions and insights NCT leads to, the same predictions could not be derived from standard evolutionary theory.

> The advocate agrees with the spirit of this comment, in the sense that the standard approach can be used to investigate any aspect of the biological world, but anticipates that the two frameworks will often lead to different predictions, and he maintains that standard neo-Darwinism does not provide *satisfactory* explanations for some phenomena. Furthermore, for the advocate the issue is not whether the same insights can be derived from the conventional perspective, but whether they actually are. Conceptual frameworks channel thinking, and the advocate believes some findings have followed more easily from the niche-construction perspective, which in itself illustrates its utility (Scott-Phillips et al. 2014: 1236).

Laland did not see niche construction theory as a replacement for standard evolutionary theory, but his reference to satisfactory explanation is similar to that of Uller and Helanterä's discussion of beach mice. This theory might provide richer accounts. However, Laland is making a second and perhaps stronger point, which is that niche construction theory might provide a framework for hypothesis generation that is more useful. This greater utility is presumably because Laland feels causation is complex and an upfront commitment to multiple causes will enable a more effective scrutiny of any and all causes of form (which is his focus). In this way, niche construction theory is seen as an idealization to deliver causal interaction accounts. This reduces niche construction theory to a heuristic role in science. But later in the same paper Laland is keen to note that niche construction is proposed as a distinct evolutionary process within a view of evolution that incorporates both variational and transformational perspectives. And yet Laland concedes that standard evolutionary theory can, as a point of principle, ask and answer all the questions niche construction theory can. In making a claim for the quasi-autonomy of niche construction theory Laland effectively reveals that niche construction theory is less an argument for the extension of the Modern Synthesis and more an attempt to incorporate older models of thinking, focused upon form. This focus upon form is Lamarckian in its developmental guise, when adopting inclusive inheritance and permitting organisms to change to fit their environment. But it is variational when focused upon the effects of organismic action upon selection pressures.

As I have hoped to demonstrate in the preceding chapters, the challenge from development fails to unseat the Modern Synthesis. The developmental niche construction literature seems to leave the fact of development stranded, in want of an explanation. It is used to make a claim about form, in the context of complaints about gene centrism, but the patterning of development and its delimited responsivity is never seen as a viable research program. I suspect this is because admitting this requirement would be to concede that niche construction theory is nothing more than an historical description of interaction in keeping with its dialectical roots.

References

Aaby BH, Ramsey G (2020) Three kinds of niche construction. Br J Philos Sci 000–000. https://doi.org/10.1093/bjps/axz054

Carneiro RL (2000) The transition from quantity to quality: a neglected causal mechanism in accounting for social evolution. Proc Natl Acad Sci USA 97(23):12926–12931. https://doi.org/10.1073/pnas.240462397
Danchin É (2013) Avatars of information: towards an inclusive evolutionary synthesis. Trends Ecol Evol 28(6):351–358. https://doi.org/10.1016/j.tree.2013.02.010
Danchin É, Pocheville A (2014) Inheritance is where physiology meets evolution. J Physiol 592 (11):2307–2317. https://doi.org/10.1113/jphysiol.2014.272096
Dawkins R (1976) The Selfish Gene. Oxford University Press, Oxford
de Regt HW (2017) Understanding scientific understanding. Oxford University Press, Oxford
Dennett DC (1983) Intentional systems in cognitive ethology: the "Panglossian paradigm" defended. Behav Brain Sci 6:343–390. https://doi.org/10.1017/S0140525X00016393
Dennett DC (1995) Darwin's dangerous idea: evolution and the meanings of life. The Penguin Press, Allen Lane
Dickins TE, Barton RA (2012) Reciprocal causation and the proximate–ultimate distinction. Biol Philos 1–10. https://doi.org/10.1007/s10539-012-9345-z
Godfrey-Smith P (1996) Complexity and the function of mind in nature. Cambridge University Press
Gould SJ, Lewontin RC (1979) The spandrels of San Marco and the Panglossian paradigm: a critique of the adaptationist programme. Proc R Soc Lond Biol Sci 205(1161):581–598. https://doi.org/10.1098/rspb.1979.0086
Griffiths PE, Gray RD (2000) Darwinism and developmental systems. In: Oyama S, Griffiths PE, Gray RD (eds) Cycles of contingency: developmental systems and evolution. MIT Press, London
Haig D (2007) Weismann Rules! OK? Epigenetics and the Lamarckian temptation. Biol Philos 22 (3):415–428. https://doi.org/10.1007/s10539-006-9033-y
Heras-Escribano M, de Jesus P (2018) Biosemiotics, the extended synthesis, and ecological information: making sense of the organism-environment relation at the cognitive level. Biosemiotics 11(2):245–262. https://doi.org/10.1007/s12304-018-9322-2
Huemer W (2019) Franz Brentano. In The Stanford Encyclopedia of Philosophy (Issue Spring) https://plato.stanford.edu/archives/spr2019/entries/brentano/
Laland KN, Odling-Smee J, Feldman MW (2004) Causing a commotion. Nature 429(6992):609. https://doi.org/10.1038/429609a
Laland KN, Sterelny K, Odling-Smee J, Hoppitt W, Uller T (2011) Cause and effect in biology revisited: is Mayr's proximate-ultimate dichotomy still useful? Science (New York, NY), 334 (6062), 1512–1516. https://doi.org/10.1126/science.1210879
Levins R, Lewontin RC (1985) The dialectical biologist. Harvard University Press, New York
Lewontin RC (1981) On constraints and adaptation. Behav Brain Sci 4(2):244–245
Lewontin RC (1983) The organism as subject and object of evolution. Scientia 118:63–82
Lewontin RC, Rose S, Kamin L (1982) Bourgeois ideology and the origins of biological determinism. Race & Class, *XXIV*(1), 1–16
Maynard Smith J (1988) Molecules are not enough. In Did Darwin get it right? Essays on games, sex and evolution (pp. 30–38). Springer. https://doi.org/10.1007/978-1-4684-7862-4_5
Maynard Smith J (2001) Reconciling Marx and Darwin. Evolution 55(7):1496–1498. https://doi.org/10.1554/0014-3820(2001)055[1496:rmad]2.0.co;2
Maynard-Smith J (1982) Evolution and the theory of games. Cambridge University Press
Mayr E (1961) Cause and effect in biology. Science 134:1501–1506. https://www.ifor.math.ethz.ch/education/bachelor/lectures/fs2011/math/Biomath/Mayr.pdf
Monod J (1971) Chance and necessity. Vintage Books
Odling-Smee FJ (1988) Niche-constructing phenotypes. In: Plotkin HC (ed) The role of behavior in evolution. The MIT Press, London, pp 73–132
Odling-Smee FJ, Laland KN, Feldman MW (2003) Niche construction: the neglected process in evolution. Princeton University Press. http://www.jstor.org/stable/j.ctt24hqpd

Plotkin HC, Odling-Smee FJ (1981) A multiple-level model of evolution and its implications for sociobiology. Behav Brain Sci 4(2):225–235. https://doi.org/10.1017/S0140525X00008566
Popper KR (1945) The open society and its enemies. Routledge
Potochnik A (2020) Idealization and many aims. Philos Sci 87(5):933–943. https://doi.org/10.1086/710622
Scott-Phillips TC, Laland KN, Shuker DM, Dickins TE, West S a (2014) The niche construction perspective: a critical appraisal. Evol Int J Org Evol 68(5):1231–1243. https://doi.org/10.1111/evo.12332
Stotz K (2017) Why developmental niche construction is not selective niche construction: and why it matters. Interface Focus 7(5):20160157. https://doi.org/10.1098/rsfs.2016.0157
Uller T, Helanterä H (2019) Niche construction and conceptual change in evolutionary biology. Br J Philos Sci 70(2):351–375. https://doi.org/10.1093/bjps/axx050
Webb RH (2011) If evolution is the answer, what is the question? J Evol Psychol 9(2):91–107. https://doi.org/10.1556/JEP.9.2011.2.2

Evolution and the Developmental Challenge

9

Abstract

This chapter draws together the view of information as systemic realization promoted throughout this book. First, Maynard Smith's analysis of the use of information in biology is discussed and I point out how his semantic version of information could lead to misinterpretation of the Modern Synthesis. The chapter then turns to an analysis of developmental systems theory, a perspective crucial to all those who have mounted a challenge from development. This helps to draw the previous chapters to a focus and to expose key differences in how theories are regarded and structured on either side of the debate. Rather than argue for pluralism due to these differences, I draw a different lesson and claim that this debate exposes a fundamental organization of biology by the Modern Synthesis as a framework that corrals multiple theories focused only upon proximate detail. I suggest that this structure is consequence of the unificatory aims of the architects of the Modern Synthesis, and that more proximate theory cannot change this fact.

Keywords

Extended Evolutionary Synthesis · Modern Synthesis · Information · Data · Context · Systemic realization · Developmental Systems Theory · Innate · Instinct · Interaction · Gene centrism · Privileging the gene · Semantic and syntactic theory · Axiom · Material inference · Framework · Corral

9.1 Introduction

Those advocating for an extended synthesis make many claims. In this book I have focused upon arguments from a developmental perspective as well as criticisms of Mayr's use of proximate and ultimate causation. From this literature it is possible to draw out three key issues or discontents with the Modern Synthesis (Gawne et al. 2018).

T. E. Dickins, *The Modern Synthesis*, Evolutionary Biology – New Perspectives on Its Development 4, https://doi.org/10.1007/978-3-030-86422-4_9

1. The exclusion of development from the Modern Synthesis
2. Gene centrism and the privileging of the gene
3. Programs of development

Concerns about the exclusion of development range from a view that developmental biology was prevented from joining the synthesis, which is unsupported by any real historical evidence, through to the concern that development was black boxed as a proximate process (Chap. 4). This last relates to the putative role of development in evolution and is strongly tied to positions taken on gene centrism and developmental programs.

The population view of evolution, that was consolidated during the synthesis, is that sorting processes such as natural selection and genetic drift cause changes in the relative frequencies of genes within a gene pool (Chaps. 2 and 3). More specifically these processes sort genetic variation, and the sources of that variation include mutation and recombination. To this end, a principal focus of the Modern Synthesis has been upon genes. Nonetheless, as Mayr was keen to argue, selection occurs at the level of the individual, the overall phenotypic expression of a particular genotype. In doing this Mayr and other Modern Synthetic thinkers were explicitly acknowledging the role of development in coordinating the production of phenotypic form, and the role of that form in selection (Chaps. 1 and 4). This is development incorporated into a variational theory of evolution.

Genes serve two roles in the preceding account. They are positioned at the initiation of development as a necessary input to that process, and they have constancy across generational time with regard to that data input function (Chap. 5). It is the property of constancy, delivered by high fidelity replication and heritability, that enables evolution. A beneficial mutation will increase in relative frequency within the gene pool as it is successfully copied. Beneficial status is determined by external contingencies. However, those contingencies are of two kinds, those external to the gene, but within the organism, and those external to the organism. The former includes interactions with other genes, and the mechanisms of development. The latter include the abiotic and biotic environment.

Developmental theorists have argued that while the gene is inherited so is much of what is external to it (Chap. 6). There is a constancy in the mechanisms of development, and a constancy in the ecological circumstances of organisms. In making these claims these theorists are arguing for an extension of the concept of inheritance beyond the gene. Thus, they claim that genetic inheritance has often been privileged over other aspects of inheritance and they seek a more inclusive model (Danchin et al. 2011).

Others argue that genes have been informationally privileged as a result of their positioning at the initiation of development (Sultan 2019). This informational privilege is due to the effects of contingency upon the selection of genetic variants, such that those variants are thought to represent a solution to environmentally posed problems. Historically this relates to Mayr's use of the term DNA program (Chap. 4) and has been associated by some critics with preformationist views (Oyama 2000). As I have argued throughout the book, the use of *information* is in need of close inspection and tight definition. I introduced a view of information as the functional relationship between data and a system, such that *information = data + context* (Chap. 5) and later expanded that discussion to draw a distinction between a view of *information as out there* to be harvested and *information as systemic realization* (Chap. 7). The idea that genes are informationally privileged belongs to the *information as out there* view. This is a more colloquial use of the term and suggests that information is a representation that captures an entire package, where a package may be a state-of-affairs or a process. For the critical developmental theorist, the Modern Synthesis saw genes as containing representational content that amounted to a blueprint or complete program for development.

The developmental arguments for extending the synthesis are arguments for pluralism when it comes to mechanisms of inheritance and to sources of representational information. In permitting more than one system of inheritance and more than one source of information they are also looking for pluralism in sources of variation, and thus the creation of form (Chap. 1). My view is that the Modern Synthesis was built in a way consistent with a *systemic realization* concept of information, and that this framework does not privilege the gene and nor does it exclude development. But what systemic realization does deliver, in the context of a population view of evolution, is a clear position on the nature of the dependencies between the causes of form.

In this final chapter I revisit information and then inspect developmental systems theory, a perspective that has influenced those seeking a developmental challenge to the Modern Synthesis. This chapter reinforces the arguments made throughout the book in defense of the Modern Synthesis and, I hope, makes the case for a systemic realization view of information at the core of evolutionary biology.

9.2 Information Revisited

Maynard Smith has defended colloquial uses of information. He begins his paper by discussing this usage as an analogy, not as a direct account of biology.

> The colloquial use of informational terms is all-pervasive in molecular biology. Transcription, translation, code, redundancy, synonymous, messenger, editing, proofreading, library—these are all technical terms in biology. I am not aware of any confusions arising because their meanings are not understood. In fact, the similarities between their meanings when referring to human communication and genetics are surprisingly close. One example must suffice. In "proofreading," the sequence of the four bases in a newly synthesized DNA strand is compared with the corresponding sequence of the old strand which acted as a template for its synthesis. If there is a "mismatch" (that is, if the base in the new strand is not complementary to that in the old strand according to the pairing rules, A-T and G-C), then it is removed and replaced by the correct base. The similarity of this process to that in which the letters in a copy are compared—in principle, one by one—with those in the original, and corrected if they differ, is obvious. It is also relevant that in describing molecular proofreading, I found it hard to avoid using the words "rule" and "correct" (Maynard-Smith 2000: 178).

Analogies are of use where there is a formal isomorphism or a qualitative similarity with the system of concern. Isomorphism is rare whereas qualitative similarity is more common, and Maynard Smith gives as his example Harvey's realization that the heart is a pump due to his exposures to pumps. Following this, and other examples, he addresses the question of whether the analogy is useful.

Maynard Smith turns to the criticism that the colloquial use relies upon a communicative notion of information and requires a coder, a decoder, a transmitter, a receiver, and an information channel. This leads Maynard Smith to ask the hard question:

> In the human example, a message is first coded, and then decoded. In the genetic case, although we think of a message in coded form in the mRNA being translated at the ribosome into the amino acid sequence of a protein, it is perhaps odd to think of this "de"-coding, since it was not "coded" from protein to mRNA in the first place. I don't think this destroys the analogy between the genetic case and the second part of the human sequence. But it does raise a hard question. If there is "information" in DNA, copied to RNA, how did it get there? (Maynard-Smith 2000: 179)

His answer is, of course, that natural selection puts the information into the DNA, and that this gives the resulting protein a meaning that can be understood as a functional role. This view is close to that defended in this book (Chap. 5). Where Maynard Smith talks of *information* being transferred and translated, I use the term *data*. Where he uses *meaning*, I use *information*. It is the data, within a specific context, that enables the *systemic realization* of information and natural selection has a hand in which contexts are preserved within a population. In this way, biological life creates information as a relational property (Boisot and Canals 2004). A key difference with Maynard Smith is that the systemic realization view strips away any commitment to semantics and focuses purely upon function. Nothing is represented by the data—it has a causal role in an overall system.

The qualitative similarity that Maynard Smith relies upon for his analogy comes from semiotic theory. He claims that codons are like symbols in their relation to amino acids in that there is no chemical reason for a particular codon to be associated with a particular amino acid, and thus the relationship is arbitrary. Given the linguistic claim that symbols permit a very large number of meanings to be communicated, Maynard Smith reasons, again by analogy, that the symbolic nature of codons enables a very large number of forms.

An arbitrary link between symbol and symbolized is indeed a standard aspect of semiotic definitions of the symbol but so is the relationship of symmetry (Dickins and Dickins 2001). Symmetry refers to the following property. The word <cat> denotes *cats*, such that upon hearing the word an English speaker can point to the appropriate organism from an array. Equally, upon having a *cat* pointed out the speaker can say <cat>. Indeed, symbols form part of an equivalence class with the object they denote and other properties of that object (Dickins 2003).[1]

We find symmetry in the complementarity of codons between DNA and RNA, but once the polypeptide chain is folded into a protein there is no symmetry between protein and DNA code. This is another way of stating the Central Dogma of Molecular Biology (Chap. 7). As we saw in Sect. 7.4.2, we cannot expect symmetrical data flow when surprisal values are unequal. Codon to codon transfers can be symmetrical as they rely upon the same alphabet and coding structure, but codons and proteins do not (Yockey 2005). Maynard Smith is relying upon the arbitrary codon → amino acid relationship to be symmetrical, but in fact most amino acids are each associated with more than one codon. There are 64 possible codons, derived from the four nucleotide bases, of which 61 represent amino acids, and the remaining three are stop codons. While the codon will correctly bring the amino acid, the amino acid will not bring the codon and so there is an asymmetry here in fact. This is redundancy and is thought to reduce the impact of incorrectly placed nucleotides in the sequence. Realizing that Shannon's theory is a theory of data helps to clarify the *data + context* relationships in this case (Floridi 2010).

All that remains of Maynard Smith's analogy is the arbitrary relationship between codons and amino acids and that is not sufficient to launch a claim for a likeness to symbolic structures. However, Maynard Smith moves onto a discussion of information as correlation, such that if A is correlated with B, it is legitimate to claim B carries information about A. This idea he borrows from Dretske (Dretske 1983) and notes that Dretske states these relations depend upon physics and local conditions, which are seen as equivalent to channel conditions for communication. This significantly weakens the symbolic claim, that relies on equivalence relations, not correlation. Instead, we can more mildly say that a gene contains information about the phenotype because we can correlate that gene with the occurrence of that phenotype. This leads Maynard Smith to draw a distinction between nature and nurture. He claims that environmental correlations, such as that between the growth rate of a baby and its nutritional environment, are different because the environment is not inherited, but genes are, and changes in evolution enact changes in nature, not nurture. This distinction leads him to introduce a key restriction in the use of information.

> I will argue that the distinction can be justified only if the concept of information is used in biology only for causes that have the property of intentionality. In biology, the statement that A carries information about B implies that A has the form it does because it carries that information. A DNA molecule has a particular sequence because it specifies a particular protein, but a cloud is not black because it predicts rain. This element of intentionality comes from natural selection (Maynard-Smith 2000: 189–190).

---

[1] Symbolic systems are also thought of as the product of convention. This in part explains the arbitrary nature of the relationship between <cat> and *cats*. In his analogy Maynard Smith is ignoring the cause of the arbitrary relationship, and he would probably be happy to call it a chemical convention as opposed to necessity.

He later adds that

> Fluctuations in the environment are a source of noise in the system, not of information. Sometimes, organisms do adapt to changes in the environment during their lifetime, without genetic evolution. For example, pigment develops in the skin of humans exposed to strong sunlight, protecting against UV. Such adaptive responses require that the genome has evolved under natural selection to cope with a varying environment. What is inherited is not the dark pigment itself, but the genetic mechanism causing it to appear in response to sunlight (Maynard-Smith 2000: 192).

The environment is thus captured as the channel conditions through which genetic information must pass. By adopting Dennett's views on intentionality Maynard Smith is effectively taking the intentional stance with regard to natural selection and treating it as if it were an agent with purposeful design intentions (Dennett 1987). Again, this is an analogical research strategy, not an ontological commitment and it enables a use of information as instruction, as representation.

In concluding, Maynard Smith returns to symbols and asks whether a protein can carry information. He distinguishes two cases. The first class are proteins that have functions because of their structure, and he gives as examples enzymes and contractile fibers. The second class are those involved in gene regulation whose functions are arbitrary with relation to their structure. What determines their functionality are specific receptor sequences in DNA, which have been naturally selected. Because protein functions are arbitrary, he argues these proteins have a symbolic, informational role. Only in this second case is there an evolved receiver. Specifically, the regulatory protein carries information for the receiver gene, and the receiver gene signals "make x here." He claims this as a semantic or intentional view of information, that is associated only with natural selection, and importantly that allows for error.

Maynard Smith sees his position as a justification for the use of information concepts in biology, and specifically of the idea that genes convey developmental information and that talk of developmental programs is justified. He does not argue for genetic blueprints. He carefully distinguishes adapted facultative responding to environmental conditions from the effects of environmental noise on information transmission and in so doing lays claim to the stability and regularity of development for genetic information and natural selection. I think his symbolic argument fails, as it misrepresents what symbols are, but if we remove this from his account, we can readily reinterpret the mechanisms he refers to in terms of *data + context* relationships. The systemic realization view has no need for the intentional stance, as it can understand persistent regularity as the outcome of selection, and error in terms of development in the absence of key inputs, sufficient input, or degraded input. There is no need to invoke pseudo-agency.

One of Oyama's key complaints about standard evolutionary approaches, such as this one, was that the metaphors and allusions introduced a fuzziness to explanation, and carried certain connotations (Oyama 2000). I have some sympathy with this view, and I think Maynard Smith's use of symbols is potentially problematic. Symbols are closely associated with linguistic and knowledge-based practices and are therefore enmeshed in intentional ascription in a way that makes it hard to adopt the intentional stance in the spirit of pragmatic science. Symbols are the arbiters of meaning in everyday life, and one reason for that is their role in equivalence classes. In its simplest rendition symbols are labels for associated classes of things—the look, sound, feel, habits, taxonomic status of a *cat* is captured by <cat>. In this way <cat> is the concept [cat] and is in turn associated with multiple possible responses, under varying contexts. Those associations are acquired knowledge (Boisot and Canals 2004). Switching a sequence of DNA on or off is not symbolic in this sense, even if the protein structure responsible is arbitrary with respect to this function. Maynard Smith also used the semiotic term *signal* and signals are either on or off to intentionally indicate presence or absence, for example. But ultimately the DNA switch is a mechanistic process in need of no observer. It is *data led*, and *context specific*. The outcome of that functional relationship is information and thinking like this prevents overwhelming our accounts with semantic interpretation that may be mistakenly considered a replacement for mechanistic theory. In this way I think that the systemic realization concept of information is a tighter analogy for biological systems, by which I mean that it is an idealization that relies on fewer, if any, untruths (Potochnik 2020).

9.3 Developmental Systems Theory

Oyama is an advocate of developmental systems theory and a key aspect of that position is the removal of dichotomies, such as that between nature and nurture. Maynard Smith uses this distinction to clarify the use of information in biology. But his dichotomy was really between natural selection and extrinsic environmental noise. His view of development is one of genetic information being used, and that information is given by natural selection. He appears to separate out non-genetic factors into those that are relevant to development, such as structural properties of proteins, and those that are not, which he labels noise. In so doing it is likely that Oyama and colleagues would see this as giving genetic information privilege in any causal account of development.

Developmental systems theory has had a distinct influence on advocates for an extended synthesis, and most especially in terms of supporting arguments for the adoption of reciprocal causation instead of the proximate–ultimate distinction proposed by Mayr (Baedke 2021; Baedke et al. 2020; Gawne et al. 2018; Mayr 1961). This is unsurprising given the focus upon the emergence of form in many of the works defending an extended view.

The origins of developmental systems theory are often placed within the behavioral sciences and in particular in Lehrman's classic critical analysis of Lorenz's concept of instincts (Johnston 2001; Lehrman 1953; Lorenz 1950). Lehrman begins his analysis by giving the example of egg retrieval in geese. An egg outside the nest is brought back in by the adult extending the neck and head forward and downward until the egg is making contact with the underside of the bill. Using the underside of the bill, the egg is rolled back and pushed between the legs of the goose. Lehrman notes the findings of Lorenz, and also Tinbergen, demonstrating that this behavior becomes exhausted upon repeated presentation of displaced eggs before the neck muscles themselves are exhausted. This led to the conclusion that the behavior was under the control of a specific neural mechanism. They also noted that if the egg escapes the goose during the retrieval behavior, the behavior continues to completion of the sequence in the absence of the egg—the adult does not stop the sequence and re-initiate it to gain the egg. Finally, the longer the latency between behavioral exhaustion and the presentation of an egg stimuli, the more likely it is that the egg-rolling sequence will restart.

Lehrman summarizes the main characteristics of this behavior that make it a case of instinct for Lorenz:

> (1) it displays a reaction-specific exhaustibility; (2) although released by stimuli coming from the egg, once released it remains constant in form regardless of variations in stimulation from the environment and even of the presence or absence of the original releasing stimulus; and (3) the threshold for elicitation falls continuously during non-exercise of the act (Lehrman 1953: 338).

Lehrman also notes that this account distinguishes the basic movement of reaching out and rolling back from any lateral movements made by the goose in order to retain the egg under the bill. It is only the movement in the sagittal plane that is regarded as an instinct, but "Lorenz maintains that every 'instinctive' behavior pattern has as its focus such an 'instinctive act' or 'consummatory act,' the performance of which serves as goal for much of the rest of the pattern" (Lehrman 1953: 338). For Lorenz, the central mechanism accumulates excitatory energy that is released upon elicitation such that the stimulus simply switches the mechanism on and the stereotyped behavioral routine is executed independently of any further stimulation. Tinbergen's concept of an innate releasing mechanism explains that the constant build-up of neural energy is inhibited, preventing continuous execution of the behavior, until such time as the appropriate stimuli is encountered. And Lorenz felt that the more energy that was accumulated then the less well formed the initial stimulation needed to be in order to elicit behavior.

Lehrman deconstructs Lorenz by recruiting empirical evidence and conceptual argument to question the distinction between innate and acquired behavior (Johnston 2001) and he concludes that the concept of instinct "depends on finalistic, preformationist conceptions of the development of behavior itself" (Lehrman 1953: 359). For Lehrman each behavior was the product of a "continuous interaction between the organism and its environment" (Johnston 2001, p.18). Johnston makes clear this did not just involve learning but also self-stimulation and other aspects of developmental experience. Where Lorenz saw pecking in chicks as stereotyped and instinctive, gradually coming online through a process of maturation, Lehrman argues that this ignores the possibility that pecking may stabilize and improve as the leg muscles and therefore balance of the chick increases.

In a later review of Thorpe's textbook Lehrman makes clear he sees the ethological use of the term innate as a form of reification and tells us that

> Thorpe assigns primary importance to the heuristic value of the distinction between "innate" and "learned," or "inherited" and "acquired," as major categories of behavior. It is the reviewer's opinion that such distinctions have value primarily when they are provisional and made for the purpose of deciding the next step in the study of development. Thorpe, however, tends to use the term innate as designating a class of behaviors for which it is possible to specify forms of organization and to suggest neural mechanisms peculiar to the class (Lehrman 1957: 104).

Here we can see Lehrman's full throated commitment to the development of all behavior, and that development would be understood in modern terms as the outcome of interactions between elements of a developmental system. In a biographical memoir of Lehrman, Rosenblatt describes him as offering an alternative to the innate, instinctive view of European ethology but from the perspective of a fellow natural historian and biologist (Rosenblatt 1995). Rosenblatt tells us that the principal difference between Lehrman and the European ethologists was that he was also well versed in "experimental methods in comparative and developmental psychology, endocrinology, and neuroendocrinology" (1995: 236) and he played a major role in bringing comparative psychologists and ethologists together. It is perhaps this disciplinary difference that led Lehrman to conclude that the terms *innate* and *learnt* were heuristic devices for the ethologists, used to organize their observations, but the further commitment to an underlying biological organization that mirrored that distinction was not warranted by the evidence of contingent development.

Lehrman revisited his 1953 criticism in a discussion of the semantic issues at the heart of nature and nurture concepts (Lehrman 1970). He was interested in semantics because he felt that the persistence of this dichotomy in behavioral biology was indicative of a dispute that could not be resolved by empirical data alone. His view on this, quoted below, is of possible relevance to the protagonists on both sides of the debate about possible extensions of the Modern Synthesis.

> When opposing groups of intelligent, highly educated, competent scientists continue over many years to disagree, and even to wrangle bitterly about an issue which they regard as important, it must sooner or later become obvious that the disagreement is not a factual one, and that it cannot be resolved by calling to the attention of the members of one group (or even of the other!) the existence of new data which will make them see the light. Further, it becomes increasingly obvious that there are no possible crucial experiments that would cause one group of antagonists to abandon their point of view in favor of that of the other group. If this is, as I believe, the case we ought to consider the roles played in this disagreement by semantic difficulties arising from concealed differences in the way different people use the same words at different times; by differences in the concepts used by different workers (i.e., in the ways in which they divide up facts into categories); and by differences in their conception of what is an important problem and what is a trivial one, or rather what is an interesting problem and what is an uninteresting one (Lehrman 1970: 18–19).

In his 1970 paper Lehrman distinguished the genetic use of inherited characters from that of some studying animal behavior. The geneticist will look to predict the distribution of a character in the offspring population, based upon its distribution in the parental population and in so doing there is no requirement to make assumptions about the developmental influences upon the focal character. He goes on to briefly discuss the concepts of heritability and epistasis and mentions Haldane's view that a genotype can exhibit more than one phenotype as a function of environmental influences. For Lehrman genes are most definitely inherited, but their functionality with reference to phenotypic outcomes can vary due to multiple factors in a developmental system. All of these comments are made against a background of assumed patterning in development.

Lehrman tells us that the contrasting view of inheritance from some studying animal behavior, and from non-geneticists, is one of *developmental fixity*. This concept of fixity is one of total resistance to environmental influence during ontogeny and, by inference, the idea is that inherited traits are already organized at the genetic level. He makes plain that this view of fixity inhabits a different conceptual world from that which allows a trait to be changed by selective breeding or natural selection. This is because an evolutionary view, while focused upon gene frequencies, is not committed to a view of how the phenotype emerges. It is not a necessary position for evolutionary theory.[2] By implication, the fixity argument, if it contacts evolutionary theory at all, adds a deterministic clause to it that is irrelevant to its central concerns about change at the population level. He accuses Lorenz and unnamed others of this error. But Lehrman is also clear that "'innate' in the sense of determination by the genome, and the concept of 'innate' in the sense of imperviousness to individual experience, refer to different problems and relationships, which cut across each other, rather than making a single conceptual whole" (Lehrman 1970: 35). In doing this he is contrasting a weak form of genetic determination of the phenotype with the strong, genetic blueprint view from Lorenz. This latter view he exposes as a weak metaphor with no basis in reality, given that blueprints are isomorphic, scaled representations of precise structure, properties that cannot be claimed for genetic data.

Lehrman's arguments were firmly directed at nativist views within behavioral biology, and he was clear to demarcate them from the practices of evolutionary biology. And yet Lehrman did not champion nurture either. As Johnston (2001) remarks, Lehrman never announced a systems theory perspective in his accounts of development. Nonetheless, he emphasized a complex interdependency between genes, learning, and experience (broadly construed) in his account of the behavioral phenotype. In so doing he set an agenda for a biology without oppositional dichotomies (Sect. 5.2.2).

Maynard Smith (2000) is not advocating developmental fixity in his discussion of information, but rather a developmental program that is a function of genetic information (data), and he would cede interaction with ontogenetic experience to deliver the final form. But that interaction is delimited by natural selection, and where it is not the environmental influences are regarded as noise. There is nothing in Lehrman's position that strictly contradicts this. His focus was upon Lorenz's

[2]Here Lehrman expresses precisely the same view of development with relation to evolution as that held by Mayr (Chap. 4). They are separate and separable processes, and theories of each have different explanatory targets.

reification of innate mechanisms, located at the neurological level.[3] Lehrman's view of behavior is not one of predetermined control, but of continuous calibration and in that sense it owes much to the view of physiological plasticity as a robustness solution (Chap. 5), while in psychological terms it is a view aligned with embodiment, or enactivism (Dickins 2021). Commitments to embodied views of cognition are not commitments to views about the role of genetic data. Indeed, the notion of calibration and responsivity requires a minimal view on purposiveness in the sense of hitting a target,[4] and that will deliver delimitation and direction. Given this the dynamic emergence of embodied behavioral control could be the product of natural selection on key feedback systems within an organism that might be energetically more parsimonious than building a central executive. This kind of heterarchical organization is not anathema to standard views within evolutionary biology (Sect. 5.2.3).

9.3.1 Defining Developmental Systems Theory

Oyama and colleagues explicitly characterize developmental systems theory as an attempt to do biology without the dichotomies of nature/nurture, genes/environment, or biology/culture (Oyama et al. 2001). The authors take aim at conventional interactionism. This position asserts that traits are formed by both nature and nurture and that we can seek to quantify how much of each for a given trait. Oyama and colleagues see this as a mistake because it perpetuates the dichotomy, attributing more-or-less causal power to one or the other half. And they claim, following Oyama's earlier work (Oyama 2000) to seek a "'stake-in-the-heart move'—a way of thinking about development that does not rely on privileged, essential causes and merely supporting or interfering causes" (Oyama et al. 2001: 1).

Inheritance and *evolution* are also lined up for reform by Oyama and colleagues, who claim that ontogeny, as a reliable process, has led to the inference that *master molecules* (by which they must mean genes) act to coordinate and determine development against a background of environmental factors that are not in want of explanation. Developmental systems theory stands in contrast to all this as a constructive process for the contingent reassembly of a great variety of resources during a life cycle. This approach can be used to define ontogeny but also evolution,

[3] Lorenz did try to save his view from Lehrman by replacing *innate* with a version of *genetic information* (Griffiths and Gray 1994). His account was semantic and misunderstood channel conditions. I suspect Maynard Smith (2000) aimed to belatedly correct Lorenz's view, or derivatives thereof, while showing how semantic interpretations might work under an intentional stance.

[4] Perceptual control theory in psychology has this structure. Its core conceit is that behavior is enacted to restore a perception. Thus, perceptions are disturbed by various factors, and organisms respond by moving, etc., in order to regain perceptual constancy (Mansell and Marken 2015). This turns behavioral explanations into closed loop accounts in contrast to standard input–output models from both behaviorism and cognitive science.

which is an ambition beyond that of Lehrman. It is the project of developmental systems theory to unravel these cycles of contingency.

Developmental systems theory "is a general theoretical perspective on development, heredity and evolution, a framework both for conducting scientific research and for understanding the broader significance of research findings" (Oyama et al. 2001: 2). Given this there are no axiomatic principles in play but instead six tenets inform the framework:

1. *Joint determination by multiple causes.* This relates directly to the commitments from Lehrman but is extended across biology.
2. *Context sensitivity and contingency.* The magnitude of effect of a single cause is a function of the state of the overall system.
3. *Extended inheritance.* Organisms inherit multiple resources that contribute to construction over the life cycle.
4. *Development as construction.* Traits are built from scratch during development and, more strongly, there is no inheritance of traits or representations of traits.
5. *Distributed control.* No component of a developmental system has control of it.
6. *Evolution as construction.* Evolution is not populations responding to environments but is change in organism–environment systems across time.

Most of these tenets are focused upon removing perceived gene centrism and gene privileging arguments from both developmental and evolutionary fields. The emphasis upon construction is referred to as an epigenetic account by Oyama and colleagues. By this they mean to oppose themselves to preformationist views and they make the interesting comment that any claim to a developmental program amounts to no more than bracketing, or black boxing those processes to return to them later, with a full account of their workings (Sect. 5.2.2). This view, combined with the general multiple causation perspective embodied in the six tenets is thoroughly proximate in its concerns, in the sense made clear by Mayr (Mayr 1961). The extension to evolutionary processes is, to all intents and purposes, a version of niche construction theory incorporating both selective and developmental perspectives on that theory (Chap. 8).

Developmental systems theory is both holistic in its approach to scientific explanation and committed to a strong version of anti-preformationism (Godfrey-Smith 2001; Griffiths 2006; Griffiths and Gray 1994). These commitments have implications for the use of information theory in biology, as Griffiths and Gray seek to demonstrate.

> In fact, there are only two ways to make sense of the notion of information in development. First, the entire set of developmental resources, plus its spatiotemporal structure, may be said to contain information about evolved developmental outcomes in the unproblematic, mathematical sense of systematic dependence. But as long as we confine ourselves to this notion of information, there is no causal asymmetry in the role of different resources which makes it legitimate to regard some of them as carrying the information and the others as merely providing conditions in which it can be read. The second, more practical way to make sense of the notion of information in development is to embed the information in one resource by

> holding the state of the other resources fixed as channel conditions under which that information is transmitted. But this move can be used to interpret any of the resources as the "seat" of the information guiding development, and so it, too, fails to generate the traditional asymmetry between genetic and other factors (Griffiths and Gray 1994: 282–283).

Griffiths and Gray are trying to show how information talk does not provide privilege for the gene. Information can flow this way and that in a biological system and there is nothing in information theory to tell us where this process might begin, or where we should start our explanations. This criticism could be levelled as systemic realization views also, in that there is nothing in that view that provides a source or direction for data flow. This position is reinforced by a criticism of Sterelny and Kitcher's background conditions argument (Sterelny and Kitcher 1988). This argument states that genes code for a trait against a background of standard genetic and environmental conditions such that if those conditions are in play any change in the phenotype can be attributed to a change in the focal gene. Griffiths and Gray point to the huge wastage of acorns that fall to the ground and rot. We are supposed to understand acorns as containing genes to build an oak, but the standard background is not one in which that happens, they claim.

> So "standard environment" cannot be interpreted statistically. The only interpretation of "standard" that will work is "such as to produce evolved developmental outcomes" or "of the sort possessed by successful ancestors." With this interpretation of "standard environment," however, we can talk with equal legitimacy of cytoplasmic or landscape features coding for traits in standard genic backgrounds. No basis has been provided for privileging the genes over other developmental resources (Griffiths and Gray 1994: 283).

All that Griffiths and Gray are claiming at this point is that there are multiple causes of the development of the phenotype, and by inference each should be attended to, and were we to start building a causal model on the foundations presented so far, we would have no principled reason to start anywhere in particular. All we are presented with is a system. But what is being assumed is that the informational talk of evolutionary theory is designed to provide that starting point and a directional steer. That is a mistake.

Maynard Smith's analysis, using a semantic take on information, made clear that it was an analogy to try and capture what evolution does, which is a more subtle use of information than Lorenz's post-Lehrman attempt to repackage his version of innateness in informational terms. Put simply, Lorenz was trying to make an ontological claim, while Maynard Smith was developing an idealization—one I have some quibbles with. In Maynard Smith's hands this was very much a natural selection led account, not untypical of the late-stage Modern Synthesis. The direction and starting points exist as they do within his analogy because that is empirically known to Maynard Smith. His case is that during the process of doing evolutionary biology and coming to understand more about how it in fact works, informational language helped to idealize the complex situation that was emerging. As an idealization this would, of course, have then enabled subsequent predictions and scientific work. And it also gave a shape to developmental processes. But the information concept was not adopted to answer these questions of direction and start. Indeed, my view is that the emergence of the Modern Synthesis led to the development of the systemic realization view, albeit tacitly, and that this is compatible with the clearest reading of Shannon type information, which is really a theory of data (Floridi 2010). The ease with which Shannon's own work was translated into colloquial information talk, and a theory of information, suggests how the Modern Synthesis might have fallen into similar practices (Floridi 2003). Indeed, a case could be made for the *information as out there* view as a form of popular idealization of both Shannon and the Modern Synthesis.

Griffiths and Gray might concede my putative historical point but quickly note that when those empirical discoveries were made, that gave Maynard Smith confidence to develop his informational analogy, they were presumably made under *an* idealization. What then, are the properties of an idealization that can solve the problem of direction and cause in such systems? This is a reasonable counter. There is no such thing as theory-less science and some model must be applied at some point, under some assumptions. But this will simply leave Griffiths and Gray with the same problem, and we will need something else to help us decide which practices to adopt. That something is, for these authors, a reconceptualization of development and evolution.

> As Oyama argued in "The Ontogeny of Information," species-typical traits are constructed by a structured set of species-typical developmental resources in a self-organizing process that does not need a central source of information. Some of these developmental resources are genetic, others, from the cytoplasmic machinery of the zygote to the social events required for human psychological development, are nongenetic. The spatiotemporal disposition of the resources is itself a critical resource, as it helps induce self-organization. The fact that appropriately structured resources are available can receive an evolutionary explanation. The processes which effectively replicate themselves are those which find appropriately structured resources in each generation. An extended notion of inheritance, which stresses the role of past generations in structuring the developmental context of their successors, is thus a critical part of the theory (Griffiths and Gray 1994: 283).

The authors make clear that this theory permits different processes but not a dichotomy of developmental resources between genes and all other resources. In making this statement they claim that some still adhere to the metaphorical idea that the other developmental resources read off what is written in the genes. It is time to move beyond this metaphor, and they suggest replacing it with that of ecological succession.

Ecological succession captures the dynamics of community ecology over time, as the colonization of a patch of land by one plant species leads to changes in soil composition and structure that enables other species to find purchase later. Those new species may displace the founders, and through their own actions lead to yet more novel species invading. All of this has implications for all the dependent trophic layers associated with colonizing plant species within the total community. These interactions and their outcomes are achieved without instruction, without a blueprint. The implication is that succession is nonetheless patterned and a useful analogy for development. As with succession, Griffiths and Gray tell us, development is about the reconstruction of form from generation to generation, using available resources, rather than the inheritance of traits. No information is inherited to guide this construction.

Ecological succession describes the changes in community structure across time as the outcome of competitive and cooperative interactions between species, in the context of changing biotic and abiotic environments. I think the intuition being primed by this analogy is that development is the coming on line of various sub-systems that constitute the overall organism, and that those sub-systems interact with one another, and have effects upon one another in much the way we would expect of a heterarchical organization (Eronen 2015). Thus, there are critical interdependencies between these sub-systems that are physically and temporally contingent. A key difference is that the sub-systems of an organism share the same DNA which give a mechanism for coordinated development, but not via direct instruction.

Dawkins' catalytic view of gene involvement in development does not rely upon semantic information but instead upon genes as necessary but not sufficient inputs to an interdependent system (Dawkins 1989). Genes can produce enzymes that act as catalysts for specific processes, and for complex products multiple catalysts can be organized into a production line sequence producing a series of intermediate forms. Each of the genes involved in each stage must act in sequence, and therefore in the context of one another as well as other developmental resources, for the product to emerge.

Single cells can cooperate to form multicellular organisms through cell specialization, which is the division of labor across the cell aggregation, such that some cells become sensory systems and others deliver circulation of the blood, etc. In multicellular organisms with specialized cells that become organs a large amount of cooperation is required to deliver this specialization and integration, much of it mediated through gene regulation. This can only happen where organisms have a bottlenecked life history, each beginning life as a single cell. It is here that Dawkins introduces a semantic, or knowledge-based analogy by referring to organisms inheriting the "*ideas* of ancestral design, in the form of the DNA program" (Dawkins 1989: 261), and using this program, development starts anew. By this he explicitly means that individuals do not inherit the same heart as one of their parents, but they instead build their own using their inherited genes to catalyze the process. Crucially, the daughter cells derived from the initial single cell will contain the same DNA, enhancing the likelihood of cooperation between them and thus successful specialization and coherence (Queller and Strassmann 2009).

Dawkins sees these integrated processes as a stereotyped growth cycle, which is the outcome of genetic switches being triggered during ontogeny and coordinating the emergence of specialized cells and a functional multicellular organism. It is this sequence that forms his DNA program, and the program is the nested operation of genes within an appropriate context.

Dawkins' view differs from that of Griffiths and Gray in one critical dimension. Dawkins uses genes, in a manner entirely consilient with a genetic data perspective, as inputs to a context that contains key resources that can be seen as a developmental system. Those inputs cause state changes, which in turn act as inputs to other systems in concert with other genetic inputs and so on. In this way genetic data has a causal and functional role in the developmental system. But there is no blueprint, no semantic instruction. Griffiths and Gray note genetic and other resources and state that they merely interact. To all intents and purposes this amounts to a high-level abstraction away from the proximate mechanics of development in precisely the same way as the concept of ecological succession is a high-level abstraction away from the cooperative and competitive interactions between individual members of various plant species. It sets loose parameters for any explanation and Dawkins' model fits comfortably within those parameters taking a sequential view on necessary conditions and their actions. But developmental systems theorists will say this sequential view privileges the gene, putting genetic data early in the sequence. That is true, but it also privileges the contexts into which those data must enter, and the systemic changes those contexts are capable of, etc. At this point, *privileging the gene* becomes nothing more than a weak mantra or banner to gather under. It is not an argument for or against a view of development.

A strong version of the abstraction argument, that I am drawn to, would be to claim there is in fact no such thing as ecological succession. More precisely there is no underlying unitary process that we can call ecological succession, but there are community level patterns of emergence due to cooperative and competitive interactions at the individual level. Equally, there is no such thing as development, or an underlying unitary developmental process, but rather interactions between single cells that lead to community level patterns. Patterned outcomes require causal accounts, and those accounts will be located at a componential level. Where we see order in the componential interactions we are within our epistemological, and probably ontological rights as evolutionary theorists, to refer to that as a program, in the sense of *a set of related activities with a shared aim* (Stevenson 2010) understood in teleonomic terms following Mayr. I suspect Griffiths and Gray are close to this view, adopting *developmental system* rather than *program* because of loose uses of the information concept in tandem with program.[5] I think this in part because Griffiths and Gray distinguish developmental outcomes that have an evolutionary explanation from those without.

[5] Program has computational connotations, and computation is generally understood as operations over stored representations by those of a classical mindset, whereas really a program is a set of operations for mediating between an input and an output. This minimal view of program clearly fits with the *data + context* view of information advocated in this book (Chap. 5). The classical view is often seen as one of transformation of input to output because of actions determined by stored representations. In this way, the stored representations carry total information to determine the outputs.

> The interactions that produce outcomes with evolutionary explanations are part of the developmental system. There is an evolutionary explanation of the fact that the authors of this paper have a thumb on each hand. We have thumbs because of the replication of thumbed ancestors. The thumb is an evolved trait. But the fact that one of us has a scar on his left hand has no such explanation. The scar is an individual trait (we are referring, of course, to the trait of having a scar just thus and so, not the general ability to scar). The resources that produced the thumbs are part of the developmental system. Some of those which produced the scar, such as the surgeon's knife, are not (Griffiths and Gray 1994: 286).

Here the thing to be explained is developmental stability across generations, and this requires an evolutionary explanation. The developmental resources that enable this, in this case, are the developmental system. The authors then tell us that

> (W)hen we privilege certain of the resources that go to construct an organism as "the developmental system" we do so to point to the explanatory connection between the transgenerational stability of these resources and the transgenerational stability of certain developmental outcomes. For other explanatory purposes, such as the study of developmental abnormalities, a different system must be delineated (Griffiths and Gray 1994: 287).

What is interesting about this statement is their use of *privilege*. Here it means that those resources that are *believed* to be directed to transgenerational developmental stability are recruited to an explanation in preference to those that are *not believed* to be so directed. Those relevant resources (made relevant presumably by processes hypothesized within an evolutionary theory) are then said to be interacting with one another, following their commitment to high-level abstraction. This is a dichotomy, equivalent to that between genetic resources and non-genetic that developmental systems theorists are eager to avoid. Furthermore, it strikes me that this is equivalent to asking questions about the reliable transmission of data across generations in the form of DNA, and its role in development, in much the way that Dawkins has. If explanatory privileging is acceptable for developmental systems theorists, why is it not acceptable for those adhering to the Modern Synthetic view of evolution? Is this just an issue of trading beliefs? *If you had the right ontology, or the right idealization, then we would be happy with you privileging whatever resources you need to just so long as those resources were listed as a part of your overall theory of x.*

I think this is the commitment and I think that developmental systems theorists hold the incorrect view that the Modern Synthesis requires a reified, semantic, and instructional view of information, when in fact it does not. This is apparent from a discussion of the distinction between two lineages of a thrush species imprinted to different habitats. Two lineages arose as a group of thrushes moved from forest to parkland and imprinted upon that, as did all subsequent generations leading to lineage isolation.[6] Griffiths and Gray claim "the genetic variation between the two populations can be presumed to be random with respect to which habitat they have imprinted on. No difference in the mechanism in the two lineages is needed to sustain their association with two very different habitats" (Griffiths and Gray 1994:

[6]This case is closely related to that of killer whales adopting different foraging habitats in Sect. 4.3

288). To fully explain the emergence of behavioral form—understood as habitat preference or choice—the incorporation of an imprinting account is required, that is not dependent upon genetic data. While imprinting mechanisms may be heritable via the transmission of genetic data across generations, the contents of the imprinted associations are a consequence of interactions with extra-genetic, and extra-organismic resources. Accordingly, these resources get packaged into the developmental system to explain the differences between these two thrush groups.

Imprinting could not happen without imprinting mechanisms, and imprinting is patterned and non-random. It does rely on stable environmental contexts to appropriately orient individuals but there are a range of contextual inputs to which imprinting mechanisms will respond, and why that range is as it is forms a standard evolutionary question.[7] *What is the mechanism for and why is it delimited as it is?* But here the assumption seems to be that the token level description could not be made sense of in terms of information in the genes—because the thrushes share the same genes—and therefore genes are insufficient explanations for that token level variation in the thrush lineages. I too think that genes are insufficient explanations for this variation, and they have never been proposed as sufficient causes, but rather necessary causes in that they contribute to the construction of imprinting mechanisms. Imprinting, like other forms of learning, is designed to afford plasticity and provide robustness for the genotype against changes in environmental contingencies during lifespan (Sect. 5.3). Certain kinds of association can be learnt, but not all.

Learning is the acquisition of data that has an informational role. Oyama has stated that development should also be seen as the creation of information (Oyama 2000). While Oyama does not explicitly commit to the systemic realization view of *data + context* as I have, it seems clear this is what she has in mind. What Oyama, and others (Griffiths and Gray 1994; Oyama et al. 2001) seem eager to do is to see evolution as a theory of form, and therefore as a theory of development and in so doing they bring a great deal of transformational apparatus into their explanations. The thrush example is a case in point. That the behavioral choice of habitat differed between thrush lineages is regarded as an issue of form, accounted for by the token level associations delivered by imprinting mechanisms that do have some underlying genetic causation. In this way ontogeny and phylogeny are brought together. But to truly make sense of the idea that ontogeny can generate information we need a theory of the nature of the proximate mechanisms of ontogeny, based in a clear understanding of data and information. And we need a clear story about the evolutionary origins of those mechanisms. That story cannot be a developmental story for logical reasons. That leaves a variational story, which is about selection and that helps to account for the current data choices made by organisms. If information is created by the

[7] Forest and parkland are in many ways very similar habitats. Indeed, traditional forests are not dense aggregations of trees, but rather scattered woodland across meadows, heath and bogland. The ecological similarity between these two habitats is probably not irrelevant for any explanation of this behavior in thrushes. Thinking like this leads to new hypotheses about the kinds of avifauna that might adopt parkland, and the kinds that might make use of more urban spaces, for example.

operation of biological systems across lifespan, it is evolution that organizes that informational capacity within the population.

If developmental systems theorists come to accept that the systemic realization view of information is at the core of the Modern Synthesis, and much of the confusion is due to colloquial usage of information, then they might be happy to privilege data in the sense of placing genetic data early in the developmental sequence, while also privileging the contexts into which that data must be inputted, etc. They might even come to heuristically use programmatic language not least because *system* can also mean "a set of principles or procedures according to which something is done; an organized scheme or method" (Stevenson 2010). That sounds programmatic.

9.4 Some Comments on Theoretical Structure

In Sect. 9.3.1 I suggested that we might argue that there is no such thing as an underlying unitary process for ecological succession nor for development. Instead, the patterning that is seen at the successional and developmental levels needs to be understood in terms of a causal account of componential interactions, the implication was that this would be couched in the language of data and context, leading to the systemic realization of information. Later I noted that Oyama, without a commitment to a theory of information of this sort, recommends seeing information as created during ontogeny, which is compatible with the position I espouse. For the remainder of that section, I argued that the developmental systems theory is not truly oppositional to the Modern Synthetic view, and indeed we could see the latter position as consilient with a developmental systems perspective when talk turned to development because the level of abstraction captured by a commitment to interaction is high and therefore epistemologically permissive. Where the two approaches differ is in how they conceive of evolutionary processes (variational versus transformational) and the developmental systems theorists' unsustainable arguments about gene centrism and privileging.

Evolution is population level change (Chaps. 2 and 3). It relies upon variation being sorted by natural selection and genetic drift. Natural selection is the statistical outcome of variation, differential success, and inheritance. Variation is proximately wrought, differential success is the outcome of biotic and abiotic interactions, and inheritance is also proximately delivered. To that end we have a causal account of population level change located in a componential layer in much the same way as succession and development might be. But this account is axiomatic in that natural selection is taken to be a truth from which further inferences and hypotheses can be drawn. A reason for this is that it is assumed that when the three components are in place within a system, selection will occur: they are necessary and sufficient conditions for this process (Webb 2011).

The axiomatic treatment of natural selection is grounded in nature by causal idealizations such as optimality modelling (Grafen 1984; Potochnik 2020). This enables the generation of empirically testable hypotheses, but crucially those hypotheses are not testing natural selection but whether a trait has been properly isolated and characterized. Where it has not it is the optimality model that is addressed, and more parameters are added, or some are removed, etc. A test of natural selection would be a test of the relationship between variation, differential success, and inheritance which is a quite different prospect. To this extent the use of optimality models to posit adaptations is a research program (Andrews et al. 2002; Gould and Lewontin 1979). At present we do not have an axiomatic theory of succession or of development and so hypotheses are created in a more pluralist fashion (Brigandt 2016; Love 2017; Moczek 2014; Travassos-Britto et al. 2021).

Love has focused upon the developmental challenge to the Modern Synthesis, which he sees as a serious attempt to incorporate explanations of form, classed as structure or morphology (Love 2017). He also makes clear that there are different kinds of *developmental form challenge*, to adopt his term, that are structured differently to one another and are not necessarily in agreement. Thus, each challenge represents a distinct idealization. His focus is upon structural issues of theory, but he is clear that much of the challenge to extend the Modern Synthesis has been grounded in a spatial metaphor (of extension) which implies incorporating more content. This is evident in Fig. 1.1 (Sect. 1.2.1) in which extra layers of content are added to the core of the synthesis. The developmental argument has predominantly focused upon the exclusion of development during the synthesis as a content claim: developmental content is missing. Love is interested in how that content might be organized within a theory.

> Philosophers have looked more narrowly at the theory structure associated with evolutionary genetics in terms of two main options: the syntactic view and the semantic view. The syntactic view argues that a scientific theory is best reconstructed as an axiom system in first-order predicate logic that includes some (small) set of exceptionless laws as core principles and an empirical interpretation of its terms. The semantic view generally treats scientific theories as families of models, where these models are indispensable to or constitutive of the meaning of the theory. Variations of the semantic view have been preferred for interpreting evolutionary theory, in part because they do not require universal laws, which appear absent from biology. However, the semantic view does not delimit the boundaries of a scientific theory and can be used to represent similar content differently (Love 2017: 166).

We should note that the syntactic view does incorporate a semantic element. The syntactic part is important because it enables the formation of axioms and then the generation of inferences from those axioms. It is first-order predicate logic that provides a language for this process. But there are non-logical terms, that describe entities and their qualities, that must be related to the logical language and can describe the core scientific aspects of the theory. These terms are semantic in that they carry veridical content about the world and are regarded as models of the underlying axiom structure (Sloep and Van Der Steen 1987). As Sloep and Van Der Steen note the syntactic view has been criticized on two main fronts—the adequacy of predicate logic for capturing theoretical structure and the adequacy of correspondence rules to map the content of non-logical terms to the world. As a result, the semantic view dispensed with syntax and simply defined theories in terms of their models, which are regarded as idealizations. Under this view a system may be defined in terms of a list of properties such that *s is an example of system S iff it contains i, ii, ...* Theoretical hypotheses are claims that some empirical system is isomorphic to the theoretical system so described. So, for example, we could reinterpret Maynard Smith's (2000) case for biological information (Sect. 9.2) as a semantic model that enables theoretical hypotheses about DNA actions, and more generally development. This conforms to Sloep and Van Der Steen's discussion about state-space versions of the semantic view in which there must be laws of state transitions and co-existence within a system. Protein synthesis, and the information flow that Maynard Smith describes are clear state transitions within development, and DNA, RNA, ribosomes, etc., are entities that can co-exist.

While semantic theories have been preferred, Love (2017) argues that they cannot tell us how content is specified or distributed to areas of evolutionary biology. This, he suggests, is an issue of scientific practices rather than an issue of syntactic or semantic characterization and this is where we should be looking to understand the emerging tensions between advocates of extension and defenders of the Modern Synthesis. This is something that Sloep and Van Der Steen also recommend, with relation to understanding evolutionary biology more generally, and they claim that it is highly unlikely "that *any* philosophical model for the reconstruction of scientific theories will be *pervasively* useful" (1987, 15).

Despite this pragmatic caution Love suggests that some evolutionary theorists do invoke a syntactic approach and that those who do so in keeping with the Modern Synthesis tend to deliver a set of axioms and non-logical terms that do not incorporate the development of form. For example, the axiomatic interpretation of natural selection, given at the beginning of this section, was tied to idealized optimality models. This is a classic syntactic interpretation, where the optimality models provide a semantic interpretation of the underlying structure. There is nothing in this structured view that will help us to explain development.

While natural selection can be treated axiomatically we should note that the theory of evolution by natural selection was not derived using first-order predicate logic, but rather it was derived principally by empirical observation coupled with the use of analogies (Chap. 2). Darwin's realization of natural selection might be better placed within propositional logic, which unlike predicate calculus does not rely upon relations and quantifiers. So, without logical notation, we can say that we have "evolution if and only if there is selection and inheritance, and we have selection if and only if there is competition and variation" (Dickins and Dickins 2018: 160). This is a form of nominalist definition and does not permit the derivation of hypotheses, in the formal sense meant by syntactic theory. In this way it captures a general Darwinian fact (Webb 2011). For formal derivation we might expect special theories, such as the Modern Synthesis, to produce their own axiomatic structures. I take it that the inclusive fitness theory is one such attempt and it is one that is in many ways still ongoing as scientists attempt to express fitness maximization mathematically and to unify it with natural selection (Grafen 2009, 2014).[8] The nominalist definition of natural selection points to a powerful empirical fact and allows that to be treated as foundational and axiomatic, but the formal work is all *post hoc*. In this way evolutionary theory does not precisely conform to the ideal of syntactic theory construction.

Love appears to interpret evolutionary developmental biology (Chaps. 6 and 7) as at an attempt to elucidate general principles of the development of form that he sees as "defeasible empirical generalizations" (Love 2017: 169). These principles can lead to hypotheses because they are linked by *material inferences*. Material inferences are not formal inferences, of the sort associated with logic, but rather are permitted by the empirical content in the premises and conclusion. In this way material inferences are often not generalizable but can nonetheless be involved in discovery (Brigandt 2010). Material inferences therefore provide structure as a function of content—not as a function of first-order predicate logic or even propositional calculus—and are strongly aligned with the semantic view by Love. Nonetheless, Love points out that paying attention to the role of material inference enables us to talk of theories, in the sense of *structures of ideas*. Thus, ecological succession and development do not require an underlying axiomatic structure to do their work, but they are successful if the conclusions drawn from the content of their premises hold empirical water. The efficacy of material inference is grounded in the practices of scientists. While Love does not directly state this, it would seem that his view is one of inductive learning such that increased exposure to certain phenomena will improve the ability to derive material inferences, but the expectation is that those inferences must be tested (Brigandt 2010): this is not a free pass for naïve inductivism.

Love concludes that developmental form challenges to the Modern Synthesis have two parts. First, they seek to incorporate proximate, mechanistic, causal models of the emergence of form into evolutionary theory. In doing this they challenge the idea that population level thinking, with its ultimate functionality is sufficient. Second, they seek to use material inference rather than axioms to structure theory. But not all developmental form challenges use the same content and they do not derive the same material inferences. Rather than seek the correct developmental challenge, Love suggests that each may be locally relevant to specific explanatory goals, but he also permits the Modern Synthesis to go about its business at the population level. His recommendation is for pluralism, with multiple theories and multiple theoretical structures.

Love's pluralism is deflationary for those seeking an extended synthesis. Rather than agree that the Modern Synthesis is conceptually inadequate and needs additional resources, he has told us that the developmental form challengers have disparate explanatory agendas with relation to one another and to those pursuing Modern Synthetic business. Moreover, the developmental form challengers have

[8] It is to inclusive fitness theory that optimality models are directly tethered. Inclusive fitness theory generally adopts the language of microeconomics in order to derive hypotheses (Parker 2006).

adopted a material inference strategy for structuring theory, while the Modern Synthesis is more closely aligned to a syntactic view. This is perhaps unsurprising given that most criticism from developmental theorists has been directed at the late stage, hardened synthesis which has been interpreted as an attempt to delineate and unify the discipline of biology (Smocovitis 1996). The quest for an axiomatic core is a method for achieving unification.

Love's characterization can be generalized to developmental systems theory. Its holism is a consequence both of multiple explanatory agendas, organized under the broad ambition of explaining form, and an avoidance of syntactic theory. Developmental systems theory does not seek axiomatic foundations, but instead is engaged in a pragmatic exercise of inductive explanation that relies on empirical testing of material inferences. The commitment to a highly abstracted notion of interaction between empirical entities is a deliberately permissive epistemology, that assumes proximate causality but not the precise nature of causal relations between components as that is something that can only be inferred under a specific question. The localized exploration of possibly relevant mechanistic causes of some aspect of form relies upon scientists being free to start their causal stories wherever they want. And as we have seen this leads to the privileging (or prioritizing) of developmental resources that are deemed relevant at the expense of others. Under this regime the idea that genes might be placed early in *all* developmental causal stories as a sufficient condition for development is anathema because that idea is an axiomatic statement from the Modern Synthesis. Technically, explanations that do this should be permitted, as argued in Sect. 9.2.1, just so long as the material inferences drawn are consistent with their conclusions and withstand empirical scrutiny.[9]

Those advancing developmental systems theory understand that the Modern Synthesis assumed selection over individuals and, given that individuals are the result of development, the fate of any genotype is dependent upon form. The models that relate the axioms of evolution to the world focus upon final forms as potentially optimal responses to contingent external facts in the world. But final form must be regarded as the outcome of developmental processes and developmental systems theorists seek to insert developmental processes into evolutionary theory at this point. Here, as we have seen in niche construction theory (Chap. 8) the idea is that the inevitable perturbations of development will influence form in ways that will introduce variance and affect evolution. The ambition of developmental systems theorists is to include perturbation as an equal cause in *evolution*, not just development. Perturbation becomes a defeasible empirical generalization when placed in the context of an inclusive inheritance theory (Chap. 7) that enables patterns of perturbation to be reliably caused across generations. This is an interesting strategy as it amounts to a confession that the axiomatic structure of natural selection is central to

[9]More precisely, individual explanations that place genes in this position should be judged on their material merits in each and every case. In doing this genes, as early sufficient causes, are not treated axiomatically. If all developmental explanations took on this structure independently there would be generality but not axiom as it was derived inductively.

evolution and thus of explanatory value. It also amounts to a confession that fidelity of transmission is a central issue for evolutionary change in populations. The distinction in thinking between those pursuing a Modern Synthetic approach and those from developmental systems theory is that the former will ask *why* the phenotype responds reliably to consistent external conditions. That *why* is Mayr's ultimate question and is the point where teleonomic engagement occurs (Chap. 4). Developmental systems theory seeks to strip this from evolutionary theory and in so doing to remove any notion of design.

The representational interpretation of information that has been assumed for the Modern Synthesis and acted as a central point of criticism is most likely a consequence of developmental systems theorists assuming information is an analogical model that helps to connect the axioms of evolution to the world. Thus, genes contain contentful information and genes are placed early in development to explain the connection between evolution and the world via form. If this is the case, then developmental systems theorists are not contrasting transformational theories with variational ones—they simply fail to see the distinction. This failure is a direct result of their focus upon local, developmental accounts of form that rely upon material inference.[10] Such local theories are nuanced and subtle proximate accounts of normative emergence within a population, but again they are not accounting for *why* the population can be normatively described in this way.

9.5 Conclusion

This chapter and this book have been focused upon criticisms of the Modern Synthesis and calls to extend it. My history of evolutionary theory has set the Modern Synthesis as a period of transition in evolutionary thinking and theory, that ended with many biologists making a strong commitment to natural selection and adaptationist thinking. But this end should not be regarded as a culmination, as the highest point of thinking achieved by evolutionary biologists. It is related to what went before, but it is also separable. The work of population geneticists does not need to use inclusive fitness theory, for example. And the game theoretic models of behavioral ecology need only make minimal and highly idealized assumptions about population genetics to derive hypotheses. In this way the connection between all aspects of evolutionary theory, as it is practiced, is simply that of natural selection understood as a process that affects gene frequencies. Those affects are caused at a distance, via the phenotypes that the genes contribute to.

I have chosen to deal with the developmental critics of the Modern Synthesis, those scholars arguing that the Modern Synthesis did not incorporate development

[10] Related to this, I suspect that material inference is very likely to be semantically driven. Anyone practicing material inference in this way is highly likely to see information talk in semantic terms and as a tool for material inference. This will cause them to interpret the Modern Synthesis as advocates of extension have.

and in so doing failed to account for form. The principal ambition of this group is to incorporate development and in so doing extend the Modern Synthesis by expanding its content (Love 2017) and its historical timeline because they believe that new findings in developmental biology must extend the thinking process (Pigliucci and Müller 2010). Embedded within this ambition are criticisms of the final phase of the Modern Synthesis—the gene was and should not have been privileged and there is no such thing as a program of development.

Gene centrism, or the privileging of the gene, is closely related to uses of information in biology. Genes have been credited with transferring information between generations, and that information is about how to build phenotypes that solve ancestral problems posed by selection pressures. This use of information has been analogical, and analogical reasoning has been integrated within evolutionary theory since Darwin's work. Darwin's analogy between artificial and natural selection was successful because there was a strong isomorphism between the two processes. But not all analogies are like this, and most rely upon similarity to some degree. Colloquial uses of information facilitate a similarity but, as I have argued, they are not formally correct definitions of what information is. Strictly, information is not something in the world to be discovered and manipulated, it is instead the functional relationship between an input and a system. It was this functionality that Shannon exploited in developing his theory of communication and that permitted him to calculate a measure of information as a measure of the value of an input within a system context.

The correct view of information, as something that is systemically realized, makes clear the intended understanding of genes within evolutionary theory. Genes, and more specifically codon sequences on DNA, are data that act as inputs into a protein synthesis system. That data has the properties of fidelity of transmission and persistence across generations that Dawkins highlighted in his work (Dawkins 1989) but it does not carry a complete set of instructions for the development of a phenotype. It carries nothing. It has functionality only within the appropriate context. To think otherwise is to commit to a form of essentialism that was anathema to the Modern Synthesis (Mayr 1982).

Dawkins (1989) has argued that development permits highly structured, multi cellular organisms to organize through cell division and differentiation. The cooperation between cells to deliver this is an outcome of evolutionary transitions and relies heavily upon gene regulation but also various other contextual factors. DNA does not direct this process, but it is data that is necessary for it. Because DNA is data it cannot be informationally privileged in the sense of containing instructions, or a blueprint for development. But because DNA data enters the developmental process as a necessary condition early and leads to systemic effects, it can be seen as privileged in this temporal manner. However, as I have made clear, in so doing we must also privilege the contexts into which DNA enters as data, that give it a functional role to play. It is the systemic nature of development, and the use of DNA data as a necessary input, that has led some to refer to DNA programs or developmental programs. This was only meant to capture systematicity, it was not a literal analogy with computation in the classical sense of operations over stored knowledge. Unfortunately, colloquial uses of information, by proponents of the Modern Synthesis, have allowed this inference to be drawn on each side of the debate. My plea has been and is for much closer inspection of informational language and its entailments.

As is apparent from these positions the Modern Synthesis did not ignore development, and the concept of the gene at work in the latter stage of the synthesis gives clear parameters to what development is and what it delivers. But the evolutionary theory of the Modern Synthesis does not deliver a precise, proximate account of the ontogeny of form. Instead, it gives an account of the kind of development that would fit with evolution as change in gene frequencies. This is a form of bracketing, leaving the detail for developmental biologists and in so doing Modern Synthetic thinkers were drawing a distinction between the processes of evolution and the process of development. This has readily been packaged into the nature/nurture dichotomy by developmental systems theorists, who are opposed to such oppositional thinking. But this dichotomy is not a necessary entailment of separating the two processes—both are aspects of nature.

The positive thesis from developmental systems theory is that evolution and development are effectively one rolling process of interaction between organisms (understood as systems) and the environment, and those interactions form a coupling between the two. This is sufficiently abstracted to make possible the inclusion of the Modern Synthetic view. Their negative thesis is that standard evolutionary approaches are preformationist because the gene is understood in terms of representational information, the kind of information that can be harvested and transmitted. Their approach to science is broadly one of semantic theory construction but borne of material inferences as they focus upon specific aspects of the development of form. Their hope is to inductively generate a mechanistic theory of form, but it may well be that the effort is more toward pluralism and multiple, bespoke mechanistic theories of forms. Nonetheless, developmental systems theorists believe that mechanistic theories of form can be incorporated into evolutionary theory as a source of stable variation in the context of the inheritance of extra-genetic factors, such as ecological variables held constant across time.

I can find nothing in the general approach of developmental systems theory that would not permit the view of development held by proponents of the Modern Synthesis. Where evolutionary developmental biologists have made very specific theoretical claims about the role of development, I find all their suggestions compatible with a view of evolutionary biology explicitly couched in the language of data. I have made the case for this in this book. Once it is realized that genes are *not* information all the core complaints from development evaporate. But, and perhaps more importantly, the interest in interaction between variables, and the multiple ways for development to unfold has proved fascinating. I have not sought to cast doubt on any empirical findings raised by developmental theorists, only upon their interpretations.

I think that the theory construction we see at work in developmental systems theory, and specifically in developmental challenges to the Modern Synthesis, is a direct consequence of the structure of the Modern Synthesis. While it is not formally axiomatic in the sense demanded by the syntactic view of theories, it can be rendered propositionally and treated as such. One consequence of the axiomatic use of evolutionary theory has been the bracketing of development in the sense of delimiting what it must be and what it is for. Those parameters are not narrow, but instead allow many possible mechanisms of development that cannot be isolated by application of the axioms of evolutionary theory. Those axioms can only later be used to check how plausible a particular developmental theory is. This necessarily leads to a discipline of development that is directed by empirical interests and local phenomena. I am clearly taking my lead from Love (2017) in these comments, but unlike him I do not see this as a move to pluralism, but rather as an organization of science. The Modern Synthesis is a framework for thinking about evolution, and it has entailments for other aspects of biology which is to be expected given its role in the unification of biology, a role that became increasingly explicit for some of its architects. As a framework it corrals the overall discipline, setting boundaries to the work of subdisciplines but it does not make precise predictions for those subdisciplines and nor does it determine methods. If there is an explicit expectation to be had it is that each area of biology will focus upon mechanistic accounts of relevant phenomena. The creation of axiomatic theory at sub-disciplinary levels is perhaps not a requirement because teleonomy has been accounted for by evolutionary theory, leaving only proximate mechanistic accounts that rely instead upon principles based in chemistry and then physics.

The concept of the Modern Synthesis as a corral might provide a more sociological insight into the discontent of developmental theorists. Love (2017) emphasizes the spatial metaphor at work in calls for extension and it might be that the Modern Synthesis feels epistemologically restrictive. This suggests a possible interpretation of Mayr's claim that developmentalists did not wish to join the synthesis: perhaps they wished to produce their own theories, unrestricted. Whatever the truth about how developmental biologists considered the emerging synthesis, the current set of arguments mounted from development so far fail to do anything more than enrich our understanding of the proximate detail of development. That is not a trivial contribution, but it does not deliver theoretical change nor even an extension to the phenomena encompassed by the Modern Synthesis.

The women and men who have crafted the developmental challenge to the Modern Synthesis are hardly fools. They are erudite and passionate advocates of their position. To echo Lehrman, I think that much of the shared miscomprehension and inability to convince is the result of semantic differences. I have laid a good portion of the blame for that with colloquial (mis)understandings of information, but I do not think that is the only source. My hope is that my survey of the current conceptual ecology has at least cleared the ground a little, perhaps exposing yet other issues to be clarified and understood. That will become labor for scientists, historians, and philosophers of science, separately and in collaboration.

References

Andrews PW, Gangestad SW, Matthews D (2002) Adaptationism – how to carry out an exaptationist program. Behav Brain Sci 25(4):489–504. http://www.ncbi.nlm.nih.gov/pubmed/12879701

Baedke J (2021) What's wrong with evolutionary causation? Acta Biotheor 69(1):79–89. https://doi.org/10.1007/s10441-020-09381-0

Baedke J, Fábregas-Tejeda A, Vergara-Silva F (2020) Does the extended evolutionary synthesis entail extended explanatory power? Biol Philos 35(1):1–22. https://doi.org/10.1007/s10539-020-9736-5

Boisot M, Canals A (2004) Data, information and knowledge: have we got it right? J Evol Econ 14 (1):43–67. https://doi.org/10.1007/s00191-003-0181-9

Brigandt I (2010) Scientific reasoning is material Inference: combining confirmation, discovery, and explanation. Int Stud Philos Sci 24(1):31–43. https://doi.org/10.1080/02698590903467101

Brigandt I (2016) Do we need a 'theory' of development? Biol Philos 31(4):603–617. https://doi.org/10.1007/s10539-015-9493-z

Danchin É, Charmantier A, Champagne FA, Mesoudi A, Pujol B, Blanchet S (2011) Beyond DNA: integrating inclusive inheritance into an extended theory of evolution. Nat Rev Genet 12 (7):475–486. https://doi.org/10.1038/nrg3028

Dawkins R (1989) The selfish gene, 2nd edn. Oxford University Press, Oxford

Dennett DC (1987) The intentional stance. MIT Press

Dickins TE (2003) General symbol machines: the first stage in the evolution of symbolic communication. Evol Psychol 1(1):147470490300100130. http://roar.uel.ac.uk/jspui/handle/10552/593

Dickins TE (2021) Ecological psychology, radical enactivism and behavior: an evolutionary perspective. Behav Philos 49:1–23

Dickins TE, Dickins DW (2001) Symbols, stimulus equivalence and the origins of language. Behav Philos 244:221–244. http://www.jstor.org/stable/10.2307/27759429

Dickins TE, Dickins BJA (2018) The extent of the modern synthesis: the foundational framework for evolutionary biology. In: Burggren W, Dubansky B (eds) Development and environment. Springer, New York, pp 155–176

Dretske FI (1983) Précis of Dretske of knowledge and the flow of information. Behav Brain Sci 6 (1):55–63

Eronen MI (2015) Levels of organization: a deflationary account. Biol Philos 30(1):39–58. https://doi.org/10.1007/s10539-014-9461-z

Floridi L (2003) From data to semantic information. Entropy 5(2):125–145. https://doi.org/10.3390/e5020125

Floridi L (2010) Information: a very short introduction. Oxford University Press

Gawne R, McKenna KZ, Nijhout HF (2018) Unmodern synthesis: developmental hierarchies and the origin of phenotypes. BioEssays 40(1):1–10. https://doi.org/10.1002/bies.201600265

Godfrey-Smith P (2001) On the status and explanatory structure of developmental systems theory. In Cycles of contingency: developmental systems and evolution (pp. 283–297). MIT Press

Gould SJ, Lewontin RC (1979) The spandrels of San Marco and the Panglossian paradigm: a critique of the adaptationist programme. Proc R Soc Lond Biol Sci 205(1161):581–598. https://doi.org/10.1098/rspb.1979.0086

Grafen A (1984) Natural selection, Kin selection and group selection. In: Krebs JR, Davies NB (eds) Behavioural ecology: an evolutionary approach. Blackwell Publishers, New York, pp 62–84

Grafen A (2009) Formalizing Darwinism and inclusive fitness theory. Philos Trans R Soc Lond Ser B Biol Sci 364(1533):3135–3141. https://doi.org/10.1098/rstb.2009.0056

Grafen A (2014) The formal darwinism project in outline: response to commentaries. Biol Philos 29 (2):281–292. https://doi.org/10.1007/s10539-013-9413-z

Griffiths PE (2006) The fearless vampire conservator: Philip Kitcher, genetic determinism, and the informational gene. In E. M. Neumann-Held, C. Rehmann-Sutter, B. H. Smith, & E. R. Weintraub (eds) (pp. 175–198). Duke University Press. https://doi.org/10.1515/9780822387336-008

Griffiths PE, Gray RD (1994) Developmental systems and evolutionary explanation. J Philos 91 (6):277–304

Johnston TD (2001) Toward a systems view of development: an appraisal of Lehrman's critique of lorenz. In: Oyama S, Griffiths PE, Gray RD (eds) Cycles of contingency: developmental systems and evolution. MIT Press, London, pp 15–23

Lehrman DS (1953) A critique of Konrad Lorenz's theory of instinctive behavior. Q Rev Biol 28 (4):337–363. https://doi.org/10.1086/399858

Lehrman DS (1957) Nurture, nature, and ethology. Contemp Psychol J Rev 2:103–104

Lehrman DS (1970) Semantic and conceptual issues in the nature-nurture problem. In: Aronson LR, Tobach E, Lehrman DS, Rosenblatt JS (eds) Development and evolution of behavior. Essays in memory of T. C. Schneirla. W.H. Freeman, London, pp 17–52

Lorenz K (1950) The comparative method in studying innate behavior patterns. Symp Soc Exp Biol 4:221–268. http://psycnet.apa.org/psycinfo/1952-05966-013

Love AC (2017) Evo-devo and the structure(s) of evolutionary theory. In: Huneman P, Walsh DM (eds) Challenging the modern synthesis: adaptation, development, and inheritance. Oxford University Press, Oxford, pp 159–187

Mansell W, Marken RS (2015) The origins and future of control theory in psychology. Rev Gen Psychol 19(4):425–430. https://doi.org/10.1037/gpr0000057

Maynard-Smith J (2000) The concept of information in biology. Philos Sci 67(2):177–194. https://doi.org/10.1017/CBO9780511778759.007

Mayr E (1961) Cause and effect in biology. Science 134:1501–1506. https://www.ifor.math.ethz.ch/education/bachelor/lectures/fs2011/math/Biomath/Mayr.pdf

Mayr E (1982) The growth of biological thought: diversity, evolution and inheritance. Belknap Press

Moczek AP (2014) Towards a theory of development through a theory of developmental evolution. In Towards a theory of development (pp. 218–226). https://doi.org/10.1093/acprof:oso/9780199671427.003.0014

Oyama S (2000) The Ontogeny of information: developmental systems and evolution, 2nd edn. Duke University Press

Oyama S, Griffiths PE, Gray RD (2001) Introduction: what is developmental systems theory? In: Oyama S, Griffiths PE, Gray RD (eds) Cycles of contingency: developmental systems and evolution. MIT Press, London, pp 1–11

Parker GA (2006) Behavioural ecology: natural history as science. In: Lucas J, Simmons L (eds) Essays in animal behaviour. Academic Press, New York, pp 23–56

Pigliucci M, Müller G (2010) Elements of an extended evolutionary synthesis. In: Pigliucci M, Müller GB (eds) Evolution: the extended synthesis. MIT Press, London, pp 3–17

Potochnik A (2020) Idealization and many aims. Philos Sci 87(5):933–943. https://doi.org/10.1086/710622

Queller DC, Strassmann JE (2009) Beyond society: the evolution of organismality. Philos Trans R Soc Lond Ser B Biol Sci 364(1533):3143–3155. https://doi.org/10.1098/rstb.2009.0095

Rosenblatt JS (1995) Daniel S. Lehrman 1919–1972. http://www.nasonline.org/publications/biographical-memoirs/memoir-pdfs/lehrman-daniel-s.pdf

Sloep PB, Van Der Steen WJ (1987) The nature of evolutionary theory: The semantic challenge. Biol Philos 2(1):1–15. https://doi.org/10.1007/BF00127559

Smocovitis VB (1996) Unifying biology: the evolutionary synthesis and evolutionary biology. Princeton University Press

Sterelny K, Kitcher P (1988) The return of the gene. J Philos 85(7):339–361

Stevenson A (ed) (2010) Oxford dictionary of english, 3rd edn. Oxford University Press, Oxford

Sultan SE (2019) Genotype-environment interation and the unscripted reaction norm. In: Uller T, Laland KN (eds) Evolutionary causation: biological and philosophical reflections. MIT Press, London, pp 109–126

Travassos-Britto B, Pardini R, El-Hani CN, Prado PI (2021) Towards a pragmatic view of theories in ecology. Oikos 1–10. https://doi.org/10.1111/oik.07314

Webb RH (2011) If evolution is the answer, what is the question? J Evol Psychol 9(2):91–107. https://doi.org/10.1556/JEP.9.2011.2.2

Yockey HP (2005) Information theory, evolution and the origin of life. Cambridge University Press, Cambridge. https://doi.org/10.1016/S0020-0255(02)00173-1

Index

T. E. Dickins, *The Modern Synthesis*, Evolutionary Biology – New Perspectives on Its Development 4, https://doi.org/10.1007/978-3-030-86422-4

The manufacturer's authorised representative in the EU is Springer Nature Customer Service Centre GmbH, Europaplatz 3, 69115 Heidelberg, Germany. If you have any concerns regarding our products, please contact ProductSafety@springernature.com

Printed and bound by CPI Group (UK) Ltd, Croydon, CR0 4YY
30/09/2026
02225695-0004